Adjutant General's Office,
June 1, 1792

By His Majesty's Command

RULES AND REGULATIONS

FOR THE

FORMATIONS, FIELD-EXERCISE

AND MOVEMENTS

OF

HIS MAJESTY's FORCES

The Naval & Military Press Ltd

published in association with

FIREPOWER
The Royal Artillery Museum
Woolwich

Published by
The Naval & Military Press Ltd
Unit 10 Ridgewood Industrial Park,
Uckfield, East Sussex,
TN22 5QE England
Tel: +44 (0) 1825 749494
Fax: +44 (0) 1825 765701
www.naval-military-press.com

in association with

FIREPOWER
The Royal Artillery Museum, Woolwich
www.firepower.org.uk

In reprinting in facsimile from the original, any imperfections are inevitably reproduced and the quality may fall short of modern type and cartographic standards.

RULES and REGULATIONS

FOR THE

FORMATIONS, FIELD-EXERCISE,

AND MOVEMENTS

OF

HIS MAJESTY's FORCES.

By His Majesty's Command.

Adjutant General's Office,
June 1, 1792.

RULES AND REGULATIONS

FOR THE

FORMATIONS, FIELD-EXERCISE,

AND MOVEMENTS,

OF

HIS MAJESTY's FORCES.

WAR-OFFICE, PRINTED;
AND SOLD BY
J. WALTER, AT HOMER's HEAD, CHARING-CROSS.
M.DCC.XCIII.

☞ ENTERED AT STATIONERS-HALL.

Adjutant General's Office,
1st June, 1792.

His Majesty thinking it highly expedient, and necessary, for the benefit of his service at large, that one uniform system of field-exercise and movement, founded on just and true principles, should be established, and invariably practised throughout his whole army, is therefore pleased to direct, that the rules, and regulations, approved of by his Majesty, for this important purpose, and now detailed and published herewith, shall be strictly followed and adhered to, without any deviation whatsoever therefrom:—And such orders before given, as may be found to interfere with, or counteract, their effect and operation, are to be considered as hereby cancelled, and annulled. It is his Majesty's farther pleasure, that the General Officers appointed to review his troops, shall be instructed to pay particular attention to the performance of every part of these Regulations, and to report their observations thereupon, for his Majesty's information, so that the exact uniformity required in all movements may be attained and preserved, and his Royal intentions thereby carried into full effect.

BY HIS MAJESTY'S COMMAND,

WILLIAM FAWCETT,
ADJUTANT GENERAL.

RULES and REGULATIONS

FOR THE

FORMATION, FIELD EXERCISE

and MOVEMENTS of

HIS MAJESTY's FORCES.

THE great object in view from the following regulations, is to establish one general and just system of movement, which directing and governing the operations of great, as well as of small bodies of troops, is to be rigidly conformed to and practised by every regiment in HIS MAJESTY's service.

To attain this important purpose, it is necessary—To reconcile celerity with order.—To prevent hurry, which must always produce confusion, loss of time, unsteadiness, irresolution, inattention to command, &c.—To ensure precision and correctness, by which alone great bodies will be able to arrive at their object in good order, and in the shortest space of time.—To inculcate and enforce the necessity of military dependance, and of mutual support in action, which are the great ends of discipline.—To simplify the execution, and to abridge the variety of movements as much as possible, by adopting such only as are necessary for combined exertions in corps, and that can be required or applied in service; regarding all matters of parade and show, meerly as secondary objects.—To ascertain to all ranks, the part each will have to act in every change of situation that can happen; so that explanation may not retard, at the moment when execution should take place.—To enable the commanding officer of any body of troops, whether great or small, to retain the whole relatively as it were in his hand and management at every instant, so as to be capable of restraining the bad effects of such ideas of independant, and individual exertion as are visionary and hurtful; and of directing them to their true and proper objects, those of order, of combined effort, and of regulated obedience, by the united force of all which, a well disciplined enemy can only be defeated.

The rules hereafter laid down will be found few, simple, and adapted to the understanding and comprehension of every individual: but they will require perfect attention in all ranks.—In the soldier an equal and cadenced march, acquired and confirmed by habit, independent of music or sound.—In the officer precision and energy of command; the preservation of just distances; and the accurate leading of divisions on given points of march, and formation.——These circumstances together with the united exertions of all, will soon attain that precision of movement, which is so essential, and without which valour alone will not avail.

These REGULATIONS are divided into PARTS, and each part subdivided into HEADS, and *Sections* of explanation.

PART I. OF THE DRILL OR INSTRUCTION OF THE RECRUIT.—— The several articles of instruction and the progression and manner in which they are to be taught, are explained in 40 *Sections*.

PART II. OF THE PLATOON OR COMPANY.——The instruction, and various operations of the company which enable it to act in battalion, are explained in 25 *Sections*.

PART. III. OF THE BATTALION.———The several operations, changes of position, and movements necessary for the battalion, when acting singly, or in line with others, are explained in 108. *Sections*.

PART IV. OF THE LINE.———The principal circumstances, relative to the movements of a considerable line are explained in 32 *Sections*

———

CONTENTS.

CONTENTS.

RECRUIT.

Without Arms.

	Sect.		Page
Each Recruit separately instructed.	1. Position of the soldier.	— — — — —	3
	2. Standing at ease.	— — — — —	4
	3. Turning eyes.	— — — — —	4
	4. The facings.	— — — — —	5
	5. Position in marching.	— — — —	6
	6. Ordinary step.	— — — —	6
	7. The halt.	— — — — —	7
	8. Oblique step.	— — — —	7
Three or four Recruits formed at open files.	9. Dressing when halted.	— — — —	8
	10. Stepping out.	— — — —	10
	11. Mark time.	— — — —	10
	12. Stepping short.	— — — —	11
	13. Changing the feet.	— — — —	11
	14. The side or closing step.	— — — —	12
	15. Back step.	— — — — —	12
	16. The quick step.	— — — — —	13

17 The

	Sect.		Page
Six or eight Recruits in one rank, at close files.	17. The quickest step.	— — — —	14
	18. File marching.	— — — —	15
	19. Wheeling a single rank, from the halt.	— — —	16
	20. Wheeling a single rank from the march.	— — —	17
	21. Wheeling a single rank backward.	— — —	18
	22. Wheeling a single rank, on a moveable pivot.	— —	18

With Arms.

	Sect.		Page
Each recruit.	23. Position of the soldier.	— — —	20
	24. Different motions of the firelock.	— — —	21
Six or Eight Files of Recruits, in a Squad.	25. Attention in forming the squad.	— — —	21
	26. Open order.	— — — — —	22
	27. Close order.	— — — —	22
	28. Manual order.	— — — —	23
	29. Platoon exercise.	— — — —	23
	30. Firings.	— — — —	23
	31. Marching to the front and rear.	— — —	23
	32. Open, and close order on the march.	— —	26
	33. March in file to a flank.	— — —	26
	34. Wheeling in file.	— — — —	27

35. Oblique

	Sect.		Page
Six or Eight Files of Recruits, in a Squad.	35. Oblique marching in front.	— — —	28
	36. Oblique marching in file.	— — — —	29
	37. Wheeling forward from the halt.	— — —	29
	38. Wheeling backward.	— — — —	29
	39. Wheeling from the march, on a halted, and moveable pivot. —		30
	40. { Stepping out, stepping short, mark time, changing feet, the side step, stepping back. — — — — — }		30

End of First Part.

Platoon or Company.

41. Formation of the Platoon. — — — — — 32
42. Marching to the front. — — — — — 34
43. The side step. — — — — — — 35
44. The back step. — — — — — — 36
45. File marching. — — — — — — 36
46. Wheeling from a halt. — — — — 37
47. Wheeling forward by sub-divisions from line. — — 37
48. Wheeling backward by sub-divisions from line. — — 38
49. Marching on an alignement in open column of sub-divisions. — — 39

50. Wheeling

Sect.		Page
50.	Wheeling into line from open column of sub-divisions.	40
51.	Sub-divisions wheeling into an alignement.	41
52.	Subdivisions wheeling into a new direction on a moveable pivot.	43
53.	Counter-march by files.	44
54.	Wheeling on the center of the platoon.	45
55.	Oblique marching.	46
56.	Increasing and diminishing the front of an open column halted.	46
57.	Increasing and diminishing on the march.	47
58.	Sub-divisions pass a defile, by breaking off files.	49
59.	Marching in quick time.	50
60.	Forming to the front from file.	50
61.	Forming from file to either flank.	51
62.	To form to either flank from column of sub-divisions.	52
63.	March in echellon by sections.	53
64.	From three ranks, forming in two ranks.	54
65.	From two ranks, forming in three ranks.	55
	Exercise of the company.	56

End of Second Part.

CONTENTS.

BATTALION.

Sect.		Page
	Formation of the Company.	1
	Formation of the Battalion,	4
	General Circumstances of Movement.	12
74.	Commands.	12
75.	Distance of files.	14
76.	Distance of ranks.	15
77.	Depth of formation.	16
78.	Music and drums.	17
79.	Marching.	18
80.	Wheeling.	20
81.	Movements.	26
82.	Points of march.	27
83.	The alignement.	28
84.	Points of formation.	29
85.	Dressing.	33
	Open Column.	36
	Battalion open Column.	38

Sect.		Page
	Assembly, &c. of the Battalion.	47
86.	Exercise of the battalion.	48
	Encrease or diminish front of Column.	55
87.	Encrease } On the march	56
88.	Diminish }	57
89.	Diminish } When halted	59
90.	Encrease }	60
	Passage of Bridge or Defile from Line.	61
91.	To the front.	61
92.	To the rear.	63
93.	Marching off in column from one flank towards another.	64
94.	March of the Battalion in File.	66
95.	General formations from file.	66

96. For-

Sect.		Page
96.	Formation in open column from file.	68

COUNTER-MARCH BY FILES. — 70

97.	Countermarch { from both flanks.	71
98.	{ from the center.	71
99.	Counter-march by divisions.	72

COUNTER-MARCH IN COLUMN. — 74

100.	Divisions by files.	74
101.	The column by divisions, from the rear.	76
102.	By wings standing, and exchanging ground.	78
103.	By wings passing thro' each other.	78

GENERAL MODES OF CHANGE OF POSITION. — 79

| 104. | The several changes of position of a battalion. | 82 |
| 105. | The several entrys on a new line, in open column. | 85 |

WHEEL AND MARCH FROM LINE INTO COLUMN, AND FROM COLUMN INTO LINE. — 90

106. 107.	{ Wheel forward into open line.	90
108. 109.	{ Wheel back into open column.	92
110.	Wheel into column of sub-divisions	94
111.	March in prolongation of the line.	94

Sect.		Page
112. 113.	{ Change of direction on a moveable pivot.	95
114.	Wheel on a halted pivot into an alignement.	95
115. 116. 117.	} Wheel of open column into an alignment.	96
118.	Halt, and wheel up of column into line.	99
119.	Wheel of sub-division column into line.	103

CHANGES OF POSITION IN OPEN COLUMN FROM LINE. — 104

120. 121.	{ On a flank halted division { to the front.	105
	{ to the rear.	106
122.	On a central division halted.	108
123.	To a distant position, by filing divisions.	109
124.	{ When the open column arrives where its head is to remain.	112
125.	{ When the open column arrives where its rear is to remain.	113
126.	{ When the open column arrives where a central division is to remain.	114
127.	{ When the open column enters the new line by the echellon march.	115
128.	{ When the divisions of the open column pass each other to form on the line	116
129.	{ When the open column forms in line by the eventail movement.	117

CHANGES

Sect.		Page	Sect.		Page
	CHANGES OF POSITION OF THE OPEN COLUMN.			DEPLOYS.	
130.	Change of position halted — on a front / on a central / on a rear division,	119	148.	On the front	136
131.			149.	On the rear — division.	137
132.			150.	On a central	138
133.	Change to a distant position — In front	120		OBLIQUE DEPLOYMENTS.	
134.	In rear.	121	151.	On an oblique line — advanced.	139
135.	To either flank.	121	152.	retired.	140
136.	Formation to flank not the pivot one.	121	153.	Formation of the line in the prolongation of the flank of the column, and on any division.	141

| | CLOSE COLUMN. | 122 |

| | ECHELLON. | 142 |

	FORMATION OF CLOSE COLUMN FROM LINE.			CHANGES OF POSITION BY COMPANY ECHELLONS.	146
137.	Before or behind a flank company.	124	154.	Wheel of battalion into echellon.	147
138.	On a central company.	125	155.	March and halt into echellon.	149
139.	March of column to a flank.	127	156.	Wheel back into parallel line.	150
140.	March of column to the front	127	157.	March up into oblique line.	151
141.	The column halted, takes a new direction.	128	158.	Formation in line from open column by the echellon march.	153
142.	The column marching, changes direction.	128	159.	The battalion { forward } on a flank	154
143.	Counter-march in close column.	129	160.	thrown { backward } company.	157
	DEPLOYMENT INTO LINE.		161.	Change of position, on a central company.	158
	CLOSE COLUMN OF COMPANIES DEPLOYS.		162.	Change of position on a distant point.	160
144.	On the front	131	163.	March of direct echellon to the front, and formation in line.	161
145.	On the rear — division.	133			
146.	On a central	134			
147.	Column of companies, forms column of two companies.	135			

ECHELLON

Sect.		Page
	ECHELLON CHANGES BY SUBDIVISIONS.	
164.	If the battalion is halted.	162
165.	If the battalion is marching in line.	165

	MARCH OF THE BATTALION IN LINE.	167
166.	When the battalion advances.	169
167.	When the battalion, halts, and dresses.	174
168.	When the battalion retires.	176
169.	Changes of direction when in movement.	178
	PASSAGE OF OBSTACLE.	181
170.	When front of obstacle is considerable.	182
171.	When the obstacle encreases.	183
172.	When the obstacle is passed, or diminishes.	184
173.	When the battalion fires in passing.	186
174.	Passing the obstacle by files.	187
175.	{ Passing a wood by companies filing.	188
	{ Passing thro' another battalion.	189

Sect.		Page
176	Retiring by alternate companies.	191
177.	{ Advancing or retiring by half battalions, and firing.	191
178.	When the battalion forms a square or oblong.	193
	March by a face.	194
	March by an angle.	
	Oblong formed from open column of march.	197
	March in open ground, prepared against cavalry.	197

End of Third Part.

Inspection or review.

Light infantry.

CONTENTS.

THE LINE.

Sec.		Page.
	MOVEMENTS OF A LINE.	2
	OPEN COLUMN OF THE LINE.	6
	GENERAL CHANGES OF POSITION OF A LINE.	9
179.	*Taking up lines of march and formation.*	13
180.	*Open column, enters, marches, forms on an alignement.*	16
181.	*Formation in line, on detached adjutants from column, or assembly in mass of battalions.*	23
182.	*Entry and formation of the rear battalions of a column, on an alignement, when the head ones have halted on it.*	29
183.	Changes of position. *On a fixed flank division.*	30
184.	*On a fixed central division.*	31
185.	*On a moving central division.*	32
186.	*Change of position, by the vourff, or quick movement.*	35

[a] CLOSE

Sec.		Page
	CLOSE COLUMN OF THE LINE.	37
187.	From colum of march, to form close column and line.	39
188.	Oblique deployments.	42
189.	When battalion close column forms square.	43
190.	Several close columns, formed from the same line.	45
191.	Several close columns form in one line.	46
192.	Two columns exchange places.	46
193.	Two columns form in one line.	47
194.	Two columns form in two lines.	47
	ECHELLON MOVEMENTS OF THE LINE.	48
195.	Oblique position taken by the echellon march.	50
196.	When from an advance in echellon, the line changes position inwards.	51
197.	When echellons advance direct from flank of line to the front.	52
198.	When a line formed on enemy's flank, attacks in echellon.	53
199.	When a line formed oblique to an enemy, attacks from a flank.	54
200.	Echellon taken from { Paralel, Oblique, Column } Position.	56
	Change of leading flank during echellon movement.	59

MARCH

Sec.		Page
	MARCH OF THE LINE IN FRONT.	59
201.	*Advance in line.*	61
202.	*Halt, and dressing of the line.*	68
203.	*Retire in line.*	70
	Lengthening the line to a flank	73
	When a line passes a bridge, or defile.	74
204.	*When a line advances or retires by half battalions.*	75
205.	*Firing in line.*	77
206.	*Square or oblong of several battalions.*	79
	CHEQUERED RETREAT OF THE LINE.	81
	Oblique position taken.	83
	Retreat of two lines.	85
	PASSAGE OF LINES.	86
	When the second line advances to relieve the first.	87
	When the second line remains posted.	87
	When a height is to be crowned.	89
	When a wing is thrown back.	89
	When a wing is refused.	90

Sec.		Page.
	SECOND LINES. — — — —	91
207.	Two lines change position on a central point of the first. —	94
208. 209.	Two lines change position {forward / backward} on a flank of the first —	96
210.	Two lines in march change {forward / backward} to a flank position by an alteration of their direction. — —	97 / 98
	COLUMN OF ROUTE. — — — —	99
	General Remarks. — — — —	106

End of Fourth Part.

PART I.

INSTRUCTION OF THE RECRUIT.

THE several heads of instruction for recruits are to be attended to, and followed, in the manner and order here set forth. It requires in the instructors to whom this duty is intrusted, and who are to be answerable for the execution of it, the most unremitting perseverance, and accurate knowledge of the part each has to teach, and a clear and concise manner of conveying his instructions; but with a firmness that will command from men a perfect attention to the directions he is giving them.—He must allow for the weak capacity of the recruit; be patient, not rigorous, where endeavour and good will are evidently not wanting: quickness is not at first to be required, it is the result of much practice. If officers and instructors are not critically exact in their own commands, and in observing the execution of what is required from others, slovenliness must take place, labour be ineffectual, and the end proposed will never be attained.

The recruit must be carried on progressively; he should comprehend one thing before he proceeds to another.—In the first circumstances of position, firelock, fingers, elbows, &c. are to be justly placed by the instructor; when recruits are more advanced, they should not be touched; but from the example shown, and the directions prescribed, be taught to correct themselves when so admonished. Recruits should not be kept too long at any particular part of their exercise, so as to fatigue or make them uneasy; and marching without arms should be much intermixed with the firelock instruction,—fife, or music, must on no account be used; but the recruit is to be confirmed by habit alone in that cadence of step which he is afterwards to maintain in his march to the enemy, in spite of every variety of noise and circumstance, that may tend to derange him.

In the manner hereafter prescribed, must each recruit be trained singly, and in squad; nor until he is steadied in these, and in other points of his duty, is he to be allowed to join the battalion; for one aukward man, imperfect in his march, or whose person is distorted, will derange his division, and of course operate on the battalion and line, in a still more consequential manner.—Every soldier on his return from long absence, must be redrilled before he is permitted to act in the ranks of his company.

Remarks upon the necessity, utility, or application, of what is hereafter prescribed, are as much as possible avoided in the first and second parts: such remarks properly belong to the third, or battalion part, with the principles of whose movements it must be supposed an instructor is sufficiently acquainted.

WITHOUT ARMS.

S. 1. Position of the Soldier.

The equal squareness of the shoulders and body to the front is the first and great principle of the position of a soldier.—The heels must be in a line, and closed.—The knees straight, without stiffness.—The toes a little turned out, so that the feet may form an angle of about 60 degrees.—Let the arms hang near the body, but not stiff, the flat part of the hand and little finger touching the thigh; the thumbs as far back as the seams of the breeches;—The elbows and shoulders to be kept back;—the belly rather drawn in, and the breast advanced, but without constraint;—the body upright, but inclining forward, so that the weight of it principally bears on the fore part of the feet;—the head to be erect, and neither turned to the right nor left.

The position in which a soldier should move, determines that in which he should stand still.—Too many methods cannot be used to supple the recruit, and banish the air of the rustic.—But that excess of setting up, which stiffens the person, and tends to throw the body backward instead of forward, is contrary to every true principle of movement, and must therefore be most carefully avoided.

N. B. The words on the margin, which are printed in *Italics*, are the words of command to be given by the instructor.

S. 2. *Standing at Ease.*

Stand at Ease. { On the words *Stand at Ease*, the right foot is to be drawn back about six inches, and the greatest part of the weight of the body brought upon it; the left knee a little bent; the hands brought together before the body; but the shoulders to be kept back, and square; the head to the front, and the whole attitude without constraint.

Attention. { On the word *Attention*, the hands are to fall smartly down the outside of the thighs; the right heel to be brought up in a line with the left; and the proper unconstrained position of a soldier immediately resumed.

When standing at ease for any considerable time in cold weather, the men may be permitted, by command, to move their limbs; but without quitting their ground, so that upon the word *Attention*, no one shall have materially lost his dressing in the line.

S. 3. *Eyes to the Right.*

Eyes Right.
Eyes Left.
Eyes Front.
{ On the words, *Eyes to the Right*, glance the eyes to the right, with the slightest turn possible of the head.—At the words, *Eyes to the Left*, cast the eyes in like manner to the left.—On the words, *Eyes to the Front*, the look, and head, are to be directly to the front; the habitual position of the soldier.

These motions are only useful on the wheeling of divisions, or when dressing is ordered after a halt; and particular attention must be paid in the several turnings

of the eyes, to prevent the foldier from moving his body, which fhould be preferved perfectly fquare to the front.

S. 4. *The Facings.*

In going through the facings, the left heel never quits the ground; the body muft rather incline forward, and the knees be kept ftraight.

1ft. Place the hollow of the right foot fmartly againft the left heel, keeping the fhoulders fquare to the front.
2d. Raife the toes, and turn to the right on both heels.
} *To the Right-face.*

1ft. Place the right heel againft the hollow of the left foot, keeping the fhoulders fquare to the front.
2d. Raife the toes, and turn to the left on both heels.
} *To the Left-face.*

1ft. Place the ball of the right toe againft the left heel, keeping the fhoulders fquare to the front.
2d. Raife the toes, and turn to the right about on both heels.
3d. Bring the right foot fmartly back in a line with the left.
} *To the Right about-face.*

1ft. Place the right heel againft the ball of the left foot, keeping the fhoulders fquare to the front.
2. Raife the toes, and turn to the left about on both heels.
3. Bring up the right foot fmartly in a line with the left.
} *To the Left about-face.*

The greateft precifion muft be obferved in thefe facings, for if they are not exactly executed, a body of men, after being properly dressed, will lose their dressing, on every small movement of facing.

S. 5. *Position in Marching.*

March. { In marching, the soldier must maintain, as much as possible, the position of the body as directed in Sect. 1. He must be well balanced on his limbs. His arms and hands, without stiffness, must be kept steady by his sides, and not suffered to vibrate. He must not be allowed to stoop forward, still less to lean back. His body must be kept square to the front, and thrown rather more forward in marching than when halted, that it may accompany the movement of the leg and thigh, which movement must spring from the haunch. The ham must be stretched, but without stiffening the knee. The toe a little pointed, and kept near the ground, so that the shoe-soles may not be visible to a person in front. The head to be kept well up, straight to the front, and the eyes not suffered to be cast down. The foot, without being drawn back, must be placed flat on the ground.

S. 6. *Ordinary Step.*

The length of each pace, from heel to heel, is 30 inches, and the recruit must be taught to take 75 of these steps in a minute, without tottering, and with perfect steadiness.

The *ordinary* step being the pace on all occasions whatever, unless greater celerity be particularly ordered, the recruit must be carefully trained, and thoroughly

instructed

instructed in this most essential part of his duty, and perfectly made to understand, that he is to maintain it for a long period of time together, both in line and in column, and in rough as well as smooth ground, which he may be required to march over. This is the slowest step which a recruit is taught, and is also applied in all movements of parade.

S. 7. *The Halt.*

On the word *Halt*, let the rear foot be brought upon a line with the advanced one, so as to finish the step which was taking when the command was given. } *Halt.*

N. B. The words *Halt, wheel—Halt, front—Halt, dress*—are each to be considered as one word of command, and no pause made betwixt the parts of their execution.

S. 8. *Oblique Step.*

When the recruit has acquired the regular length and cadence of the ordinary pace, he is to be taught the oblique step. At the words, *To the Left, Oblique.—March*, without altering his personal squareness of position, he will, when he is to step with his left foot, point, and carry it forward 19 inches in the diagonal line, to the left, which gives about 13 inches to the side, and about 13 inches to the front. On the word *Two*, he will bring his right foot 30 inches forward, so that the right heel be placed 13 inches directly before the left one. In this position he will pause, and on the word *Two*, continue to march, as before directed, by advancing his left foot 30 inches, pausing at each step till confirmed in his

To the Left, Oblique March.

Fig. 1.

this position; it being essentially necessary to take the greatest care that his shoulders be preserved square to the front. From the combination of these two movements, the general obliquity gained will amount to an angle of about 25 degrees. When the recruit is habituated to the lengths and directions of the step, he must be made to continue the march, without pausing, with firmness, and in the cadence of the ordinary pace, viz. 75 steps in the minute.

As all marching (the side-step excepted) invariably begins with the left foot; whether the obliquing commences from the halt, or on the march, the first diagonal step taken is by the leading foot of the side inclined to, when it comes to its turn, after the command is pronounced.

The squareness of the person, and the habitual cadenced step, in consequence, are the great directions of the oblique, as well as of the direct march.

Each recruit should be separately and carefully instructed in the principles of the foregoing eight sections of the drill. They form the basis of all military movements.

———

Three or four recruits will now be formed in one rank, at very open files, and instructed as follows.

S. 9. *Dressing when halted.*

Dress. Dressing is to be taught equally by the left as by the right. On the word *Dress*, each individual will cast his eyes to the point to which he is ordered to dress, with the smallest turn possible of the head, but

preserving

preserving the shoulders and body square to their front. The whole person of the man must move as may be necessary, and bending backward or forward is not to be permitted. He must take short quick steps, thereby gradually and exactly to gain his position, and on no account be suffered to attempt it by any sudden or violent alteration, which must infallibly derange whatever is beyond him. The faces of the men, and not their breasts or feet, are the line of dressing. Each man is to be able just to distinguish the lower part of the face of the second man beyond him.

In dressing, the eyes of the men are always turned to the officer, who gives the word *Dress*; and who is posted at the point by which the body halts; and who from that point corrects his men, on a point at, or beyond his opposite flank.

The faults to be avoided, and generally committed by the soldier in dressing, are, passing the line; the head too forward, and body kept back; the shoulders not square; the head turned too much.

Two, or more men, being moved forward, or backward, a given number of paces, and placed in the new line, and direction, the following commands will be given.

> By the *Right, forward—Dress.*
> By the *Right, backward—Dress.*
> By the *Left, forward—Dress.*
> By the *Left, backward—Dress.*

As soon as the dressing is accomplished, the words *Eyes Front*, will be given, that heads may be replaced, and remain square to the front. } *Eyes front.*

No rank, or body, ought ever to be dressed, without the person on its flank appointed to dress it, determining, or at least supposing a line, on which the rank, or body, is to be formed, and for that purpose taking as his object the distant flank man, or a point beyond such flank, or a man thrown out on purpose;—dressing must then be made gradually, and progressively, from the fixed point, towards the distant flank one; and each man successively, but quickly, must be brought up into the true line, so as to become a new point, from whence the person directing proceeds in the correction of the others; and he himself, when so directing, must take care that his person, or his eyes at least, be in the true line, which he is then giving.

S. 10. *Stepping out.*

Step out. { The squad marches as already directed in ordinary time. On the word *Step out*, the recruit must be taught to lengthen his step to 33 inches, by leaning forward a little, but without altering the cadence.

This step is necessary when a temporary exertion in line, and to the front, is required; and is applied both to ordinary and quick time.

S. 11. *Mark Time.*

Mark Time. { On the word, *Mark Time*, the foot then advancing compleats its pace; after which the cadence is continued, without gaining any ground, but alternately throwing out the foot, and bringing it back square with
Ordinary Step. { the other.—At the word *Ordinary Step*, the usual pace of 30 inches will be taken.

This

This ſtep is neceſſary marching in line, when any particular battalion is advanced, and has to wait for the coming up of others.

S. 12. *Stepping Short.*

On the word, *Step Short*, the foot advancing will finiſh its pace, and afterwards each recruit will ſtep as far as the ball of his toe, and no farther, until the word *Ordinary Step*, be given, when the uſual pace of 30 inches is to be taken. } *Step Short. Ordinary Step.*

This ſtep is uſeful when a momentary retardment of either a battalion in line, or of a diviſion in column, ſhall be required.

S. 13. *Changing the Feet.*

To change the feet in marching, the advancing foot compleats its pace, the ball of the other is brought up quickly to the heel of the advanced one, which inſtantly makes another ſtep forward, ſo that the cadence may not be loſt. } *Change Feet.*

This may be required of an individual, who is ſtepping with a different foot from the reſt of his diviſion; in doing which he will in fact take two ſucceſſive ſteps with the ſame foot.

S. 14. *The Side or Closing Step.*

The side step is performed from the halt in ordinary time, by the following commands.

Close to the Right——March.
Close to the Left——March.

Close to the Right, March.

Halt.

> In closing to the right, on the word *March*, eyes are turned to the right, and each man carries his right foot about 12 inches directly to his right (or if the files are closed, to his neighbour's left foot), and instantly brings up his left foot, till the heel touches his right heel; he then pauses, so as to perform this **movement in ordinary time**, and proceeds to take the next step in the same manner; the whole with perfect precision of time, shoulders kept square, knees not bent, and in the true line on which the body is formed.—At the word *Halt*, the whole halt turn their eyes to the front, and are perfectly steady. (Vide S. 43.)

S. 15. *Back Step.*

Step back, March.

Halt.

> The *Back Step* is performed in the ordinary time and length of pace, from the halt, on the command *Step back,—March.*—The recruit must be taught to move straight to the rear, preserving his shoulders square to the front, and his body erect.—On the word *Halt*, the foot in front must be brought back square with the other.

A few paces only of the back step can be necessary at a time.

S. 16.

S. 16. *The Quick Step.*

The cadence of the ordinary pace having become perfectly habitual to the recruits, they are now to be taught to march a quick time, which is 108 steps in the minute, each of 30 inches, making 270 feet in a minute.—The command *Quick, March*, being given with a pause between them; the word *Quick*, is to be considered as a caution, and the whole to remain perfectly still and steady; on the word *March*, they step off with the left feet, keeping the body in the same posture, and the shoulders square to the front; the foot to be lifted off the ground, that it may clear any stones, or other impediments in the way, and to be thrown forward, and placed firm; the whole of the sole to touch the ground, and not the heel alone; the knees are not to be bent, neither are they to be stiffened, so as to occasion fatigue, or constraint.—The arms to hang with ease down the outside of the thigh; a very small motion to prevent constraint may be permitted; but not to swing out, and thereby occasion the least turn, or movement of the shoulder; the head is to be kept to the front, the body well up, and the utmost steadiness to be preserved.

Quick, March.

This is the pace to be used in all *filings* of divisions from line into column, or from column into line; and by battalion columns of manœuvre, when independently changing position.—It may occasionally be used in the column of march of small bodies, when the route is smooth, and no obstacles occur; but in the march in line of a considerable body it is not to be required, and very seldom in a column of manœuvre; otherwise fatigue must arise to the soldier, and more time will be lost by hurry and inaccuracy, than is attempted to be gained by quickness.

The

The word *March*, given fingly, at all times denotes that *ordinary* time is to be taken; when the *quick march* is meant, that word will precede the other.—The word *March* marks the beginning of movements from the *halt*; but is not given when the body is in previous motion.

S. 17. *The Quickeſt Step.*

The *quickeſt time*, or *wheeling march*, is 120 ſteps of 30 inches each, or 300 feet in the minute.—The directions already given for the march in quick time relate equally to the march in quickeſt time.

This is applied chiefly to the purpoſe of wheeling, and is the rate at which all bodies accompliſh their *wheels*, the outward file ſtepping 33 inches, whether the wheel is from line into column, during the march in column, or from column into line.—In this *time* alſo ſhould diviſions double, and move up, when paſſing obſtacles in line; or when in the column of march, the front of diviſions is encreaſed, or diminiſhed.

Three or four recruits in one rank, with intervals of 12 inches between them, ſhould be practiſed in the different ſteps, that they may acquire a firmneſs and independence of movement.

Many different times of march muſt not be required of the ſoldier.—Theſe three muſt ſuffice, ORDINARY TIME (75 ſteps in the minute), QUICK TIME (108 in the minute), WHEELING, or QUICKEST TIME (120 in the minute).

PLUMMETS,

PLUMMETS, which vibrate the required times of march in a minute, are of great utility, and can alone prevent or correct uncertainty of movement; they must be in the possession of, and constantly referred to by each instructor of a squad,— the several lengths of plummets swinging the times of the different marches in a minute are as follows:

		In.	Hund.
Ordinary time, - - -	75 steps in the minute	24	96
Quick time, - - - -	108 - - - -	12	03
Quickest, or wheeling time, -	120 - - - -	9	80

A musket ball suspended by a string which is not subject to stretch, and on which are marked the different required lengths, will answer the above purpose, may be easily acquired, and should be frequently compared with an accurate standard in the adjutant's, or serjeant-major's possession. The length of the Plummet is to be measured from the point of suspension to the center of the ball.

Accurate distances of steps must also be marked out on the ground, along which the soldier should be practised to march, and thereby acquire the just length of pace.

———————

Six or eight recruits will now be formed in a rank, at close files, having a steady, well-drilled soldier on their flank to lead,—and FILE MARCHING may be taught them.

S. 18. *File Marching.*

The recruits must first *face,* and then be instructed to cover each other exactly in file, so that the head of the man immediately before, may } *To the—face.*

may conceal the heads of all the others in his front.—The strictest observance of all the rules for marching is particularly necessary in marching by files, which is first to be taught at the *ordinary time*, and afterwards in *quick time*.

March.

On the word *March*, the whole are immediately to step off together, gaining at the very first step 30 inches, and so continuing each step without encreasing the distance betwixt each recruit, every man locking or placing his advanced foot on the ground, before the spot from whence his preceding man had taken up his,—no looking down, nor leaning backward is to be suffered, on any pretence whatever,—the leader is to be directed to march straight forward to some distant object given him for that purpose, and the recruits made to cover one another during the march, with the most scrupulous exactness,—great attention must be paid to prevent them from marching with their knees bent, which they will be very apt to do at first, from an apprehension of treading upon the heels of those before them.

S. 19. *Wheeling of a single Rank, in ordinary Time from the Halt.*

Right Wheel.

March.

At the word, *To the Right wheel*, the man on the right of the rank faces to the right; on the word *March*, they step off together, the whole turning their eyes to the left (the wheeling flank), except the man on the left of the rank, who looks inwards; and, during the wheel, becomes a kind of base line for the others to conform to, and maintain the uniformity of front.—The outward wheeling man always lengthens his step to 33 inches,—the whole observe the same time, but each man shortening his step in proportion as he is nearer to the standing flank on which

which the wheel is made,—during the wheel, the whole remain closed to the standing flank; that is, they touch, without incommoding their neighbour; nor must they stoop forward, but remain upright,—opening out from the standing flank, is to be avoided; closing in upon it, during the wheel, is to be resisted.—On the word *Halt, Dress,* each man halts immediately, without jumping forward, or making any false movements.   *Halt, Dress.*

When the recruits are able to perform the wheel with accuracy in the *ordinary time,* they must be practised in wheeling in *quickest time.*

Nothing will tend sooner to enable the recruit to acquire the proper length of step, according to his distance from the pivot, than continuing the wheel without halting for several revolutions of the circle.—And also giving the word *Halt, Dress,* at instants not expected, and when only a 6th, 8th, or any smaller proportion of the circle is compleated.

S. 20. *Wheeling of a single Rank, from the March.*

The recruits are first to be taught to perform this wheeling at the *ordinary time,* and afterwards in the *quickest,* or *proper wheeling time,*—the rank, marching to the front at the ordinary time, receives the word of command, *Halt, Right, Wheel,* the man on the right of the rank   *Halt, Right, Wheel.* instantly halts, and faces to his right; the rest of the rank, turning their eyes to the wheeling flank (as directed in the preceding section), immediately change the step together to *wheeling time;* as soon as the portion of the circle to be wheeled is completed, the words *Halt, Dress,*   *Halt, Dress.* will be given, (a pause of 2 or 3 seconds may be made), and then *March,*

March. { *March*, on which the whole rank steps off together at the ordinary time.

S. 21. *Wheeling Backwards, a single Rank.*

On the Right backwards, Wheel.
March.

{ At the word *On the Right backwards, Wheel*, the man on the right of the rank faces to his left; at the word *March*, the whole step backward in wheeling time, dressing by the outward wheeling man, those nearest the pivot man making their steps extremely small, and those towards the wheeling man encreasing them as they are placed nearer to him.—The recruit in this wheel must not bend forward, nor be suffered to look down; but by casting his eyes to the wheeling flank, preserve the dressing of the rank.—On the word *Halt*, the whole remain perfectly steady, still looking to the wheeling flank till they receive the word *Right Dress*.

Halt.

Right Dress.

The recruits should be first practised to wheel backwards at the ordinary step; and at all times it will be necessary to prevent them from hurrying the pace; an error soldiers are very liable to fall into, particularly in wheeling backwards.—Where large bodies wheel from line into column, this wheeling is necessary to preserve the covering of pivot flanks, and the distances of the divisions, which the line is to break into.

S. 22. *Wheeling of a single Rank on a moveable Pivot.*

In wheeling on a moveable pivot, both flanks are moveable, and describe concentric circles round a point, which is removed a few paces from what would otherwise

wise be the standing flank; and eyes are all turned towards the directing pivot man, whether he is on the outward-flank, or on the flank wheeled to.

When the wheel is to be made to the directing pivot flank, (suppose the left)—the rank marching at the ordinary pace, receives the word, *Right Shoulders forward*; on which the pivot man, without altering either the time or length of his pace, continues his march on the circumference of the lesser circle, and tracing out a considerable arch, on the principle of dressing, gradually brings round his rank to the direction required, without obliging the other flank, which is describing the circumference of a larger circle to too great hurry;—on the word *Forward*, shoulders are squared, and the pivot marches direct to his front. *Right Shoulders forward.*

Forward.

When the directing pivot is on the outward flank, and has to describe the circumference of the larger circle, on the word *Left shoulders, forward*, he will, without changing the time, or length of his pace, gradually bring round the rank to the required direction, so as to enable the inward flank to describe a similar arc of a lesser circle, concentric to the one he himself is moving on.—During both these wheels, the rank dresses to the proper pivot, and when he describes the smaller circle of the wheel, the other flank which has more ground to go over, will quicken its march, and step out.—When the pivot describes the greater circle of the wheel, the other flank, which has less ground to go over, will step shorter, and gradually conform——In the first case, the recruit must be cautioned against opening out from the pivot; and, in the latter, from crowding on him. *Left Shoulders forward.*

The just performance of this mode of wheeling depends so much on the directing pivot, that a well drilled soldier should, at first, be placed on the flank named, as the proper pivot, and changed occasionally.—It is used, when a column of march (in order to follow the windings of its route), changes its direction in general, less than the quarter circle.

WITH ARMS.

S. 23. *Position of the Soldier.*

When the firelock is given, and is shouldered, the person of the soldier remains in the position described (Section I.) except, that the wrist of the left hand is turned out, the better to embrace the butt, the thumb alone is to appear in front, the four fingers to be under the butt, the left elbow is a little bent inwards, without being separated from the body, or being more backward or forward than the right one.——The firelock is placed in the hand, not on the middle of the fingers, and carried in such manner that it shall not raise, advance, or keep back one shoulder more than the other; the but must therefore be forward, and as low as can be permitted without constraint; the fore part nearly even with that of the thigh, and the hind part of it pressed by the wrist against the thigh; the piece must be kept steady and firm before the hollow of the shoulder; should the firelock be drawn back or attempted to be carried high, in that case, one shoulder will be advanced, the other kept back, and the upper part of the body distorted, and not placed square with respect to the limbs.

Each recruit must be separately taught the position of shouldered arms, and not allowed to proceed until he has acquired it.

S. 24.

S. 24. *Different Motions of the Firelock.*

The following motions of the firelock will be taught and practised as here set down, until each recruit is perfect in them; they being necessary for the ease of the soldier in the course of exercise.

As mentioned in the manual exercise.
- Supporting arms.
- Carrying arms.
- Ordering arms.
- Standing at ease.
- Attention.
- Shouldering from the order.

The recruit must be accustomed to *carry* his arms for a considerable time together; it is most essential he should do so, and not be allowed to support *them* so often as is practised, under the idea of that long *carrying* them is a position of too much constraint.

A platoon, company, or battalion, are never to MARCH, or HALT, or FORM IN LINE, or to DRESS, (which are situations where the greatest accuracy of front is required) but with *carried Arms.*—When such bodies are standing and halted, arms may be occasionally *supported.*—When marching in column, or that small divisions are moving any distance in file, firelocks may also be *supported.*

S. 25. *Attention in forming the Squad.*

When the SQUAD or division (consisting of from six to eight files) is ordered to *fall in,* each man with carried arms, will as quick as possible take his place in his rank,

rank, beginning from the flank, to which he is ordered to form; he will dress himself in line by the rule already given; assume the ordered position of a soldier, and stand perfectly still, and steady, until ordered to stand at ease, or that some other command be given him.—Attention must be paid, that the files are correctly closed; that the men in the rear ranks cover well, looking their file leaders in the middle of the neck;—That the ranks have their proper distance of one pace (30 inches) from each other;—That all the ranks are equally well dressed;—That the men do not turn their heads to the right, or left; and that each man has the proper unconstrained attitude of a soldier.

S. 26. *Open Order.*

Rear ranks take Open Order.

March.

The recruits being formed in three ranks at close order, on the word *Rear ranks take Open Order*, the flank men on the right and left of the centre, and rear ranks, step briskly back one and two paces respectively, face to their right, and stand covered, to mark the ground on which each rank is to halt, and dress at open order; every other individual remains ready to move.—On the word *March*, the dressers front, and the center and rear ranks fall back one and two paces, each dressing by the right the instant it arrives on the ground.

S. 27. *Close Order.*

Rear ranks take Close Order.

March.

On the word *Rear ranks take Close Order*, the whole remain perfectly steady; at the word *March*, the ranks close within one pace, marching one and two paces, and then halting.

S. 28.

S. 28. *Manual Exercise.*

According to Regulation.

S. 29. *Platoon Exercise.*

According to Regulation.

S. 30. *Firings.*

When the recruits have acquired the management of their arms, and are perfect in the motions of the manual, and platoon exercises, they will be instructed at closed ranks in firing.

> Direct to their front.
> Obliquely to the right and left.
> By files.

S. 31. *Marching to the Front and Rear.*

The squad, or division, is to be particularly well dressed; files correct; arms carried; the rear ranks covering exactly, and each individual to have his just attitude and position before the squad is ordered to move.—The march will be made by the right or left flank, and a proper trained man will therefore conduct it.—The word *Squad*, or *Division*, may be given as a caution; and at the word *March*, each man steps forward a full } *Fig. 2.*

Caution.
March.

full pace.—The recruit muſt not turn his head to the hand to which he is dreſſing, as a turning of the ſhoulders would undoubtedly follow.— His elbows muſt be kept ſteady, without conſtraint; if they are opened from his body, the next man muſt be preſſed upon; if they are cloſed, there ariſes an improper diſtance which muſt be filled up; in either caſe waving on the march will take place, and muſt therefore be avoided.

Halt, front.
March.
Turning to the right or left, or about, in march, is not to be at firſt practiſed; but the ſquad is to *halt, front* by command, and then *march.*

On many occaſions where a body great or ſmall after a movement to the rear, or in file, is immediately to reſume its proper front; inſtead of the words to halt, and face about, the word *Halt front*, as one command, will be given, when it is inſtantly to face to its proper front in line.—Nor in general ſhould there be any ſenſible pauſe between the halt front of any body, and it is after fronting, that the dreſſing if neceſſary, is ordered to take place.

As the being able to march ſtraight forward is of the utmoſt conſequence, he who commands at the drill will take the greateſt pains in making his ſquad do ſo;— For this purpoſe he will often go behind his ſquad, or diviſion, place himſelf behind the flank file by which the ſquad is to move in marching, and take a point, or object, exactly in front of that file; he will then command *March*, and remaining in his place, he will direct the advance of the ſquad, by keeping the flank file always in a line with the object.—It is alſo from behind, that one ſooneſt perceives the leaning back of the ſoldier, and the bringing forward or falling back of a ſhoulder; faults which ought inſtantly to be rectified, as productive of the worſt conſequence in a line, where one man, by bringing forward a ſhoulder, may change the direction of the march, and oblige the wing of a battalion to run, in order to keep dreſſed.

In short, it is impossible to labour too much at making the soldier march straight forward, keeping always the same front as when he set off.—This is effected by moving solely from the haunches, keeping the body steady, the shoulders square, and the head to the front; and will without difficulty be attained by a strict attention to the rules given for marching, and a careful observance of an equal length of step, and an equal cadence, or time of march.

Changing from *ordinary* to *quick time*, and from *quick* to *ordinary time*, must always be preceded by a previous, but instantaneous *halt*; although this may not appear essential for the movements of a squad, division, or battalion, it is absolutely so for those of a larger body, and is therefore required in small ones.

Turning on the march, in order to continue it, though inaccurate and improper for a large body, is necessary, when companies, or their divisions are moving in file, and that without halting it is eligible to make them move on in front; or when moving in front, it is proper without halting to make them move on in file. } *Right Turn.* *Left Turn.*

As helps for fixing the true time, or cadence of the march, the plummet must be frequently resorted to; the words *left*, *right*, may when necessary be repeated, slowly for ordinary time, and quicker for quick time.—Strong taps of the drum, if in just time, and regulated by the plummet, are also directed to be given immediately before the word *March*, thereby to imprint the required measure on the mind of the recruit; but they are on no account, or in any situation, to be given during the march.

S. 32. *Open, and Close Order, on the March.*

Rear Ranks take Open Order. { The squad, when moving to the front in ordinary time, receives the word *Rear ranks take Open Order*, on which the front rank continues its march, without altering the pace, and the center and rear ranks mark the time, viz. the center once, and steps off at the second step; the rear rank stepping off on the third pace.

Rear Ranks take Close Order. { On the word *Rear ranks take Close Order*, the centre and rear ranks step nimbly up to close order, and instantly resume the pace, at which the front rank has continued to march.

S. 33. *March in File to a Flank.*

The accuracy of the march in file is so essential in all deployments into line, and in the internal movements of the divisions of the battalion, that the soldier cannot be too much exercised to it.—The whole battalion, as well as its divisions, is required to make this flank movement without the least opening out, or lengthening of the file, and in perfect cadence, and equality of step.

To the—face. March. { After *facing*, and at the word *March*, the whole squad steps off at the same instant, each replacing, or rather over-stepping the foot of the man before him; that is the right foot of the second man comes within the left foot of the first, and thus of every one, more or less over-lapping, according to the closeness, or openness of the files, and the length of step.—The front rank will march straight along the given line, each soldier of that rank must look along the

necks

necks of thofe before him, and never to right or left; otherwife a waving of the march will take place, and of courfe the lofs, and extenfion of line, and diftance, whenever the body returns to its proper front. —The center and rear ranks muft look to, and regulate themfelves by, their leaders of the front rank, and always drefs in their file.—Although file marching is in general made in quick time; yet it muft alfo be practifed, and made in ordinary time. The fame pofition of feet, as above, takes place in all marching in front, where the ranks are clofe, and locked up.

With a little attention and practice this mode of marching, which appears fo difficult, will be found by every foldier to be eafier than the common method of marching by files, when on every halt the rear muft run up to gain the ground it has unneceffarily loft.

S. 34. *Wheeling in File.*

The fquad, when marching in file, muft be accuftomed to wheel its head to either flank; each file following fucceffively, without lofing, or encreafing diftance.—On this oecafion, each file makes its feparate wheel on a pivot moveable in a very fmall degree, but without altering its time of march, or the eyes of the rear ranks being turned from their front rank.—The front rank men, whether they are pivot men, or not, muft keep up to their diftance, and the wheeling men muft take a very extended ftep, and lofe no time in moving on.

The head of a company or battalion marching in file, muft change direction in the fame manner on the moveable pivot, by gradually gaining the new from the old direction, and thereby avoiding the fudden ftop that otherwife would take place.

S. 35. *Oblique Marching in Front.*

Right Oblique.

Forward.

> When the Squad is marching in front, and receives the word *To the Right, oblique*; each man, the firſt time he raiſes the right foot, will, inſtead of throwing it ſtraight forward, carry it in the diagonal direction, as has been already explained in Sect. 8. taking care not to alter the poſition of his body, ſhoulders, or head.—The greateſt attention is to be paid to the ſhoulders of every man in the ſquad, that they remain parallel to the line on which they firſt were placed, and that the right ſhoulders do not fall to the rear, which they are very apt to do in obliquing to the right, and which immediately changes the direction of the front.—On the word *forward*, the incline ceaſes, and the whole march forward.—In obliquing to the left, the ſame rules are to be obſerved, with the difference of the left leg going to the left, and attention to keeping up the left ſhoulder.

The ſame inſtructions that are given for ordinary time, ſerve alſo for quick time; but this movement, though it may be made by a ſquad, or diviſion, cannot be required from a larger body in quick time.

Obliquing to the right, is to be practiſed ſometimes with the eyes to the left; and obliquing to the left, with the eyes to the right; as being abſolutely neceſſary on many occaſions;—for if one of the battalions of a line in advancing be ordered to oblique to the right, or to the left, the eyes muſt ſtill continue turned towards its center.

S. 36.

S. 36. *Oblique Marching in File.*

In obliquing to the right, or left, by files, the center and rear rank men will continue looking to their leaders of the front rank.—Each file is to confider itself as an entire rank, and is to preferve the fame front, and pofition of the fhoulders, during the oblique, as before it began.—This being a very ufeful movement, the recruits are to be often practifed in it.

S. 37. *Wheeling forward from the Halt.*

The directions already given for the wheeling of a fingle rank (vide Sect. 19.) are to be ftrictly attended to in this wheel of the fquad.— On the word *Right (or left) Wheel*, the rear ranks, if at one pace diftance, lock up. At the word *March*, the whole ftep together in the quickeft time, and the rear ranks during the wheel, incline fo as to cover their proper front rank men.—At the word *Halt*, the whole remain perfectly fteady.

Right Wheel.

March.

S. 38. *Wheeling backward.*

The Squad muft be much practifed in wheeling backward in the quickeft time.— In this wheel, the rear ranks may preferve their diftance of one pace from each other.—Great attention fhould be paid, to prevent the recruits from fixing their eyes on the ground. (Vide Sect. 21.)

S. 39.

S. 39. *Wheeling from the March, on a halted, and moveable Pivot.*

The directions for wheeling on a *halted*, and on a *moveable* pivot, have already been given, in Sects. 20, and 22.——The squad should now be practised in both, until the recruits are thoroughly confirmed in those movements.

S. 40. *Stepping out,—Stepping short.—Marking the Time,— Changing the Feet,—The Side Step,—Stepping back.*

{ The squad must likewise be practised in, *stepping out, stepping short marking the time, changing the feet, the side step*, and *stepping back*, the instructions for which have been fully detailed in the foregoing sections.

It cannot be too strongly inculcated, or too often recollected, that upon the correct *equality* of *march*, established and practised by all the troops of the same army, every just movement and manoeuvre depends. When this is not attended to, disunion, and confusion, must necessarily take place, on the junction of several battalions in corps, although, when taken individually, each may be, in most respects well trained: It is in the original instruction of the recruit, and squad, that this great point is to be laboured at, and attained.—The *time* and *length* of step, on all occasions, are prescribed. The TIME is infallibly ascertained, by the frequent corrections of the *plummet*, which, when so applied, will soon give to each man that habitual measure so much desired; and therefore every driller must have it constantly in his hand; and, as it has been already observed, before any squad, or larger body, is put in march, 5 or 6 strong taps of the drum should often be given

in

in exact time, as regulated by the plummet; which will imprint the true meafure on each ear, and prepare for taking an accurate ftep at the word *March*. The length of ftep is only to be acquired by repeated trial, and therefore, before the recruit, or fquad, is put in motion, each inftructor fhould afcertain the fpace on which he is to drill his men; he will therefore (fuppofing that he himfelf is accurate in his paces, and that there is ground for that purpofe) mark out an oblong fquare of 40 paces by 20, or 30, the corners of which he will afcertain by halberts, ftones, or in any other vifible manner; along the fides of this figure he will march the pivot flank of the fquad, making correct wheels, and halts at the angles.—The time of March being fo exactly afcertained, he will then fee that the fides of the oblong are gone over at the known number of fteps; and if there be any inaccuracy, he will lengthen or fhorten the ftep, till the fquad marches with the utmoft precifion; every man preferving his juft pofition, and all the other indifpenfible attentions in marching being ftrictly obferved.—Where there is a fufficiency of *ground*, the fquads will occafionally march over greater fpaces, but the diftances fhould in the fame manner be exactly afcertained, fo that there may be no doubt as to the true length of the ftep.—In proportion to the ftrength of fquads, or drills, one or more formed foldiers fhould accompany each, to march on the flank, give diftances, and in other points, to regulate the motions of the drill.

D. D.

End of PART FIRST.

PART II.

OF THE PLATOON, OR COMPANY.

S. 41. *Formation of the Platoon.*

The recruit being thoroughly grounded in all the preceding parts of the drill, is now to be inftructed in the movements of the platoon, as a more immediate preparation for his joining the battalion: for this purpofe from 10 to 20 files are to be affembled, formed, and told off in the following manner, as a company in the battalion.

The platoon FALLS IN, in three ranks at clofe order, with fhouldered firelocks; the files lightly touching, but without crouding; each man will then occupy a fpace of about 22 inches.—The commander of the platoon takes poft on the right of the front rank, covered by a ferjeant in the rear rank.—Two other ferjeants will form a fourth or fupernumerary rank, three paces from the rear rank.

The platoon will be told off into fub-divifions, and if of fufficient ftrength, into four fections; but as a fection fhould never be lefs than five files, it will often happen that for the purpofes of march, three fections only can be formed.

The four best trained soldiers are to be placed in the front rank, on the right and left of each sub-division.

When thus formed, the platoon will be practised in

Opening, and Closing of } Ranks.   (Sec. 26 and 27.)

Dressing { to the front, to the rear, in an oblique direction, } by the right and left;

and be exercised in the several motions of the firelock, as have been shewn in the preceding part.

Close order is the chief and primary order in which the battalion, and its parts, at all time assemble, and form.—*Open order* is only regarded as an exception from it, and occasionally used in situations of parade, and show.—In close order, the rear ranks are closed up to within one pace; the length of which is to be taken from the heels of one rank to the heels of the next rank.—In open order, they are two paces distant from each other.

{ In order to distinguish the words of command given by the instructor of the drill who represents the commander of the battalion), from those given by the commander of the platoon, or its divisions, the commands of the former are in CAPITAL Letters, those of the latter in *Italic*.

S. 42. *Marching to the Front.*

Fig. 2.

MARCH.

In the drill of the platoon, the person instructing must always consider it as a company in battalion, and regulate all its movements upon that principle; he will therefore, before he puts it in motion to front, or rear, indicate which flank is to direct, by giving the word EYES RIGHT, or EYES LEFT; and then MARCH.— Should the right be the directing flank, the commander of the platoon himself will fix on objects to march upon in a line truly perpendicular to the front of the platoon; and when the left flank is ordered to direct, he and his covering serjeant will shift to the left of the front rank, and take such objects to march upon.— To MARCH on one object only, and to preserve a straight line, is an operation not to be depended on; the conductor of the platoon before the word MARCH is given, will therefore endeavour to remark some distinct object on the ground, in his own front, and perpendicular to the directing flank: he will then observe some nearer and intermediate point in the same line, such as a stone, tuft of grass, &c. these he will move upon with accuracy, and as he approaches the nearest of those points, he must from time to time chuse fresh ones in the original direction, which he will by this means preserve, never having fewer than two such points to move upon. If no object in the true line can be ascertained, his own squareness of person must determine the direction of the march

A person placed in the rear of a body can, more readily than if placed in its front, determine the line which is perpendicular to such front; and could we suppose ranks and files most perfectly correct, the prolongation of each file would be a perpendicular to the front of the body.

As the MARCH of every body, except in the case of inclining, is made on lines perpendicular to its then front, each individual composing that body must in his

person

perfon be placed, and remain perfectly fquare to the given line; otherwife he will naturally and infenfibly move in a direction perpendicular to his own perfon, and thereby open out, or clofe in, according to the manner in which he is turned from the true point of his March.—If the diftortion of a fingle man operates in this manner, and all turnings of the head do fo diftort him, it may be eafily imagined what that of feveral will occafion, each of whom is marching on a different front, and whofe lines of direction are croffing each other.

Accuracy and fquarenefs of pofition, the equality of cadence and ftep, the light touch of the files, which is never to be relinquifhed, juft diftances, and true lines of momement, will give, without apparent conftraint, the head being turned, or the leaft trouble taken in dreffing, the moft decifive exactnefs in the marches, and operations of the largeft bodies.

The platoon, during its march in line, will occafionally be ordered to

Step out	vide Sect.	10
Mark time	———	11
Step fhort	———	12
Open, and *clofe ranks*	———	32
Oblique	———	35

S. 43. *The Side Step.*

The *fide*, or *clofing ftep*, muft alfo be frequently practifed; it is very neceffary and ufeful on many occafions, when halted, and when a very fmall diftance is to be moved to either flank:—As for inftance, to open, or clofe files; to join one divifion to, or open it from another; to regain an interval in line; to move a whole battalion, or parade, 20 or 30 paces to a flank; to regulate distances between close columns before deploying:—alterations made in this manner are imperceptible from the front, and better made than by facing, and file marching: the words of command must be decided and strong.

To the right close. March. Halt.	When the whole platoon is to close, at the word To the right close, the platoon officer takes one step to the front and instantly faces about, the covering serjeant replacing him: On the word March, the whole move together agreeably to the directions (in Sect. 14). On the word Halt, the platoon officer resumes his place, having stepped in the same manner as the men, but fronting them, and thereby assisted in preserving the direction.

S. 44. *The Back Step.*

Step back—March.	The platoon must be accustomed from the halt, at the words Step back—March, to step back any ordered number of paces in the ordinary time and length, as it is an operation that may be frequently required from a battalion.

S. 45. *File Marching.*

Left face. Quick march.	In marching by files, the commander of the platoon will lead the front rank; therefore when the movement is by the left, on the word To the left face, he, and his covering serjeant, will instantly shift to the left flank of the platoon; at the word Quick march, the whole

steps

steps off together, (vide sect. 18); and on the word *Halt, Front*, the leader, and his serjeant, will return to their posts on the right. — *Halt, Front.*

---

46. *Wheeling from a Halt.*

In wheeling either forward, or backward from a halt, the commander of the platoon, on the word RIGHT or LEFT WHEEL, moves out, and places himself one pace in front of the center of his platoon: during the wheel, he turns towards his men, and inclines towards that flank which has been named as the directing, or pivot one, giving the word *Halt, Dress*, when his wheeling man has just compleated the required degree of wheel: he then squares his platoon, but without moving what was the standing flank, and takes his post on the directing flank. — RIGHT WHEEL. MARCH. *Halt, Dress.*

S. 47. *Wheeling forward by Sub-divisions from Line.*

On the Caution BY SUBDIVISIONS, TO THE RIGHT WHEEL, the commander of the platoon places himself one pace in front of the center of the right sub-division, at the same time the men on the right of the front rank of each sub-division face to the right. — CAUTION.

At the word MARCH, each sub-division steps off in wheeling time, observing the directions given in (Sect. 19 and 37). The commander of the platoon turning towards the men of the leading sub division, and inclining to its left (the proper pivot flank), gives the word *Halt, Dress*, for both sub-divisions, as his wheeling man is taking the last step that — MARCH. Fig. 3. A. *Halt, Dress.*

that finishes the wheel square; and instantly posts himself on the left, the pivot flank.—The serjeant coverer, during the wheel goes round by the rear, and takes post on the pivot flank of the second sub-division.—It is to be observed, that the commander of the platoon invariably takes post with the leading sub-division; therefore, when the platoon wheels by sub-divisions to the left, the commander of the platoon moves out to the center of the left sub-division, and during the wheel inclines towards the right, now become the proper pivot flanks of the sub-divisions.

The *proper* pivot flank is column is that which, when wheeled up to, preserves the divisions of the line in the natural order, and to their proper front: the other may be called the *reverse* flank.

In column, divisions cover and dress to the proper pivot flank: to the left when the right is in front: and to the right when the left is in front.

S. 48. *Wheeling backward by Sub-divisions from Line.*

CAUTION.
{ The platoon will also break into open column of sub-divisions by wheeling backwards.—When the right is intended to be in front; at the caution BY SUBDIVISIONS ON THE LEFT, BACKWARD WHEEL, the commander of the platoon moves out briskly and places himself in front of the center of the right sub-division.—The man on the left of the front rank of each sub-division at the same time faces to the right.

MARCH.
{ On the word MARCH, each sub-division wheels backward in quickest time, as directed in Sect. 21, and Sect. 38. During the wheel, the
commander

commander of the platoon turns towards his men, inclining at the same time to the left, or pivot flank, and on compleating the wheel, gives the word *Halt, Dress*, to both divisions: he, and his covering serjeant, then place themselves on the left flanks of their sub-divisions. } Fig. 4. **A.** *Halt, Dress.*

It may be considered as a rule almost general (the reasons for which are given in the following part) that all wheels of the battalion, or line, (when halted and when the divisions do not exceed 16, or 18 files,) into column, should be backward.—And all wheels from column into line, forward.—The only necessary exceptions seem to be in narrow ground where there is not room for such wheels.

S. 49. *Marching on an Alignement, in Open Column of Sub-divisions.*

The platoon having wheeled backwards by sub-divisions from line, (as directed in the foregoing Section) and a distant marked object in the prolongation of the two pivot flanks being taken; the commander of the platoon, who is now on the pivot flank of the leading sub-division, immediately fixes on his intermediate points to march on, (vide Sect. 42.) On the word MARCH, given by the instructor of the drill, both divisions step off at the same instant; the leader of the first division marching with the utmost steadiness and equality of pace on the points he has taken; and the commander of the second division preserving the leader of the first in an exact line with the distant object; at the same time he keeps the distance necessary for forming from the preceding division; which distance is to be taken from the front rank.—These objects are in themselves sufficient to occupy the whole attention of the leaders of the two divisions; therefore they must not look to, nor endeavour to correct the march of their divisions, which care must be entirely left to the non-commissioned officers of the supernumerary rank. Fig. 4. **A.**

S. 50.

S. 50. *Wheeling into Line from Open Column of Sub-divisions.*

HALT.	The platoon being in open column of sub-divisions, marching at the ordinary step on the alignement, receives the word HALT, from the instructor of the drill; both divisions instantly halt, and the instructor sees that the leaders of the divisions are correct on the line in which they have moved; he then gives the word (supposing the right of the platoon to be in front) by sub-divisions TO THE LEFT WHEEL INTO LINE; on which the commander of the platoon goes to the center of his sub-division, the two pivot men face to their left exactly square with the alignement, and a serjeant runs out and places himself in a line with them, so as to mark the precise point at which the right flank of the leading sub-division is to halt, when it shall have compleated its wheel.— At the word MARCH, the whole wheel up in quickest time; during the wheel, the commander of the platoon, turning towards his men, inclines to the wheeling flank, and gives the word *Halt, Dress*, at the moment the wheel of the division is compleating; the commander of the platoon, if necessary, corrects the internal dressing of the platoon on the serjeant and pivot men; this dressing must be quickly made, and when done, the commander of the platoon gives the word *Eyes front*, in a moderate tone of voice, and takes post in line as directed in Sect. 41.
LEFT WHEEL INTO LINE.	
MARCH.	
Halt, Dress.	
Eyes front.	

In all wheels of the divisions of a column that are to be made on a halted pivot in order to form line, the flank firelock of the front rank on the hand wheeled to, is such pivot, not the officer who may be on that flank, and whose business it is to conform to it.

All wheelings by sub-divisions, or sections, from line into column, or from column

lumn into line, are performed on the word given by the commander of a battalion, when the whole of a battalion is at the same instant so to wheel, or on the word given by the commander of the company, when companies singly, or successively, so wheel: they are not to be repeated by the leaders of its divisions.

S. 51. *In Open Column of Sub-divisions wheeling into an Alignment.*

The platoon being in open column of sub-divisions, marching in ordinary time; when its leading division arrives at the ground, where the wheel is to commence, it receives the word *Halt, right,* or *left, wheel,* from its commander; on which the rear ranks, if at one pace distance, lock up; the flank front rank man alone halts, and faces into the new direction, while the others quicken their pace to the wheeling time, and regulate their step by the outward hand (to which they have turned their eyes), until the wheel is compleated.—He then gives the word *Halt, Dress,* for his division to dress to the hand it is to move by; and whenever the second division, which has continued to advance in ordinary time, arrives close on the wheeling point, he gives his division the word *March,* and moves on in ordinary time, so as its rear rank does not occasion even a momentary stop to the division behind it, which at that instant receives the word *Wheel,* then *Halt, Dress,* and finally *March,* whenever the leading division has gained its proper distance from it.

Halt, Wheel.

Halt, Dress.

March.

Halt, Wheel.
Halt, Dress.
March.

The officer conducting the leading (and every other) division of the column in march, on any given point or object where it is to wheel into a new direction, and to its proper pivot hand on a halted pivot; always stops at that point, or object, close on his own outward hand, and gives the word WHEEL, when the front

f

rank

rank of his division has taken ONE pace beyond such object; he thus allows space for his own person (when the wheel is finished) to move on close behind the new direction of march.

But if the proper pivot flank is to be the wheeling one, each commander of a division gives his word *Wheel,* as he successively arrives at such a distance from the point on which he has moved, as that at the completion of the wheel, his division may *halt* perpendicular to the new line, but with the given point, of course, behind the proper pivot; and that he also in his own person be on the new direction, prepared to give his word *March,* and to proceed.

The sub-divisions must take care that they continue their march correctly upon, and wheel exactly at the point where the leading one wheeled, and that they do not shift to either flank, which without much attention they are apt to do.

In this manner the sub-divisions succeed each other; and if the words of command be justly given; no stop made on arriving at the wheeling point; the wheels performed at an increased time and step; and the proper halt, dressing, and pause, be made after the wheel; no extension of the column will take place, but the just distances between the divisions will be preserved.

The officer conducting the directing flank of a division may during the wheel be advanced one or two paces before it, and remain so, facing to the flank, that he may the more critically be enabled to give his word *Halt*; at which instant, he will again place himself on the flank ready to judge his distance, and to give the word *March.*

S. 52. *In Open Column of Sub-divisions wheeling into a new Direction, on a moveable Pivot.*

The commander of the leading sub-division, when at a due distance from the intended new direction, will give the word *Right* (or left) *Shoulders forward* (vide Sect. 22), and he himself carefully preserving the rate of march, without the least alteration of step or time, will begin to circle in his own person from the old into the new direction, so as not to make an abrupt wheel, or that either flank shall be stationary; the rest of his division on the principle of dressing will conform to the direction he is giving them: when this is effected he will give the word *Forward.*—The leader of the second sub-division, when he arrives at the ground on which the first began to wheel, will in this manner follow the exact tract of the first, always preserving his proper distance from him.

Right Shoulders forward.

Forward.

Thus without the constraint of formal wheels; a column, when not confined on its flanks, may be conducted in all kinds of winding and changeable directions; for if the changes be made gradual and circling, and that the pivot leaders of divisions pursue their proper path at the same uniform equal pace, the true distances of divisions will be preserved, which is the great regulating object on this occasion, and to which every other consideration must give way.

To which ever hand the wheel is made on a moveable pivot, it is made within and cuts off the angle formed by the intersection of the old and new directions.

In wheeling in column of march on a fixed pivot, the outward file whether officer or man is the one wheeled on.

S. 53. *Countermarch by Files.*

The platoon, when it is to countermarch, must always be considered as a division of a battalion in column; the instructor of the drill will therefore, previous to his giving the caution to countermarch, signify whether the right or left is supposed to be in front, that the commander of the platoon, and his covering serjeant, may be placed on the pivot flank before such caution is given, as it is an invariable rule in the countermarch of the divisions of a column by files, that the facings be made from the flank, then the pivot one, to the one which is to become such.

FACE.	On the word, To THE RIGHT, or left, FACE, the platoon faces, the commander of it immediately goes to the other flank, and his covering serjeant advancing to the spot which he has quitted, faces to the right about.—At the word QUICK MARCH, the whole, except the serjeant coverer, step off together, the platoon officer wheeling short round the rear rank (viz. to his right, if he has shifted to the right of the platoon; or to his left, if he be on the left of it); and proceeds, followed by the platoon in file, till he has conducted his pivot front rank man close to his serjeant, who has remained immoveable; he then gives the words *Halt, Front*, and *Dress*, squares, and closes his platoon on his serjeant, and then replaces him.
QUICK MARCH.	
Halt, Front. Dress.	

All countermarches by files necessarily tend to an extension of the files; unity of step is therefore absolutely indispensible, and the greatest care must be taken that the wheel of each file be made close, quick, and at an increased length of step of the wheeling man, so as not to retard or lengthen out the march of the whole.

Companies, or their divisions when brought up in file to a new line, are not to stand

ſtand in that poſition, till the men cover each other minutely; but the inſtant the leading man is at his point, they will receive the word *Halt, front,* and in that ſituation cloſe in, and dreſs correctly.

S. 54. *Wheeling on the Center of the Platoon.*

The platoon muſt be accuſtomed to wheel upon its center, half backward, half forward, and to be pliable into every ſhape, which circumſtances can require of it; but always in order, and by a decided command.

The Words of Command are,

PLATOON, ON YOUR CEN- TER TO THE { RIGHT, LEFT, RIGHT ABOUT, LEFT ABOUT, } WHEEL.

When the wheel to be made is to the right, or right about, the right half platoon is the one to wheel backward, and the left forward.— The reverſe will take place, when the wheel is to be made to the left, or to the left about.—On the word MARCH, the whole move together in the quickeſt time, regulating by the two flank men, who during the wheel preſerve themſelves in a line with the center of the platoon;— as ſoon as the required degree of wheel is performed, the commander of the platoon gives the word *Halt, Dreſs,* and inſtantly ſquares it from that flank, on which he himſelf is to take poſt.

MARCH.

Halt, Dreſs.

S. 55.

S. 55. *Oblique Marching.*

The instructor of the drill will have the oblique march frequently practised, in platoon, in sub-divisions, and in file: (Vide Sect. 35. 36.) He will see when in divisions, that the rear ranks lock well up, and cover exactly;—when in file, that the exact distances are preserved between the files;—and in both cases, that the platoon during its march, continues parallel to the position, from which it *commenced* obliquing.

S. 56. *Increasing and diminishing the front of an open column halted.*

Fig. 5. B.   *Increasing.*

FORM PLATOON.

Rear S. division, left Oblique.
Q. March.
Forward.
Halt, dress.

The platoon standing in open column of subdivisions (suppose the right in front) receives from the instructor of the drill a caution to Form Platoon——The commander of the platoon turning round instantly orders, *Rear Subdivision, left Oblique.—Q. March.* When it has obliqued so as to open its right flank, he gives the word *Forward;* and on its arriving in a line with the first division he orders, *Halt, Dress,* and takes post on the left, the pivot flank of the platoon.

Fig. 5. A.   *Diminishing.*

FORM SUBDIVISIONS.

Left Subdivision inwards face.

On the cautionary command from the instructor of the drill to FORM SUBDIVISIONS, the serjeant coverer falls back to mark the point where the left flank of the subdivision is to be placed.—The commander of the platoon advancing one step, orders *Left Subdivision, inwards face,* and instantly on facing, the three leading files disengage to the rear.

At

At the word *Q. March*, the file passes round, and behind the serjeant, and at the proper instant receives the words; *Halt, front—Left, dress.—* The commander of the platoon is now on the left flank of the first sub-division, and his serjeant on that of the second.

Q. March.
Halt, front.
Left, dress.

It is to be observed as a general rule in diminishing the front of a column, by the doubling of sub-divisions or sections (whether the column be halted or in motion) that the sub-division or section, on the *reverse* flank, is the one behind which the other sub-division, or sections double.——Thus when the right is in front, the doubling will be in the rear of the right division; and, vice versa, when the left is in front; by which means, the column is at all times in a situation to form line to the flank, with its divisions in their natural order, by simply wheeling up on the pivot flanks.—And in encreasing the front of a column, the rear sub-divisions, or sections, oblique to the hand the pivot flank is on; so that when the right is in front, the obliquing will be to the left; and the reverse when the left is in front.

S. 57. *Increasing and diminishing the Front of an Open Column on the March.*

Increasing.

Fig. 6. B.

The platoon marching at the ordinary time in open column of sub-divisions (suppose the right in front), receives from the instructor of the drill, the cautionary command, FORM PLATOON; the commander of the platoon instantly turning round gives the words, *Left oblique—quick march*; on which the rear sub-division obliques to the left, and as soon as its right flank is open, receives the word, *Forward.*—When it gets up to the first sub-division (which has continued to march, with the utmost steadiness,

FORM PLATOON.
Left oblique.
Quick march.
Forward.

Ordinary. { steadiness, at the ordinary pace), the commander of the platoon gives the word *Ordinary*, and takes post on the pivot flank, towards which he has been moving.

Fig. 6. A.

Diminishing.

FORM SUB-DIVISIONS.

Left Sub-division.
Mark time.

Quick Oblique.

Forward.

{ When the instructor of the drill gives the caution to FORM SUB-DIVISIONS, the commander of the platoon advancing one step, immediately orders, *Left sub-division, mark time;* this it does until the right one, which continues its march steadily at the ordinary pace, has cleared its flank; he then orders the left sub-division, *Quick oblique,* and when he perceives that it has doubled properly behind the right one, he gives the word, *Forward,* on which it takes up the ordinary march, and follows at its due distance of wheeling, he himself being then placed on the pivot flank of the sub-division, and his serjeant on that of the second.

The same directions that apply to encreasing or diminishing by sub-divisions, apply equally by sections, which individually repeat the same operations.

Increasing and reducing the front of a column, is an operation that will frequently occur in the march of large bodies; and it is of the utmost importance that it be performed with exactness.—The instructor of the drill must therefore be particularly attentive, that the transition from one situation to the other be made as quick as possible; that the leading division continues its march at the regular time, and length of pace, and the exact distances between the divisions be accurately preserved.—During the operation, the ranks must be closed, arms carried, and the greatest attention required from each individual.

———————

S. 58.

S. 58. *The Platoon in Open Column of Sub-Divisions to pass a short Defile, by breaking off Files.*

The platoon is supposed in open column of sub-divisions, with the right in front, marching in ordinary time; when the leading division is arrived within a few paces of the defile, it receives from the instructor of the drill an order to break off a certain number of files, (suppose three).—The commander of the leading division instantly gives the words, *Three files on the left, right turn;* the named files immediately turn to their right, and wheel out in rear of the three adjoining files.— The commander of the sub-division himself closes into the flank of the part formed.—When the second sub-division comes to the spot where the first division contracted its front, it will receive the same words of command from its own leader, and will proceed in like manner.

Fig. 7. C.

BREAK OFF 3 FILES.

Three files, right turn.

Should it be required to diminish the front of the column one or two files more, the commander of the leading division will, as before, order the desired number of files to *turn;* on which those already in the rear will incline to their right, so as to cover the files now ordered to break off, and which are wheeling out in the manner already prescribed.

Two files, right turn.

In this movement, the files in the rear of the sub-divisions must lock well up, so as not to impede the march of the succeeding division.

As the defilè widens (or the instructor of the drill shall direct) the commander of the leading sub-division, will order files to move up to the front, by giving the word, *One, two,* or *three files to the front;* on which the named files turn to their front (the left), and lengthening their

Three files to the front.

their pace, march up, file by file, to the front of their fub-divifion, and immediately refume the ordinary pace.—Thofe files which are to continue in the rear will oblique to the left, lengthning alfo their ftep, till they cover, and are clofed up to the three files on the left flank of their fub-divifion.

S. 59. *Marching in Quick Time.*

The platoon muft frequently be practifed to march in quick time, particularly in file, until the men have acquired the utmoft precifion in this movement, which is fo effential in all deployments from clofe column.—The platoon will alfo occafionally be marched in front at the fame ftep, as it may be frequently required from fmall bodies.

S. 60. *Forming to the Front from File.*

HALT, FRONT.
CAUTION.

MARCH.

The platoon when marching in file may form to its front, either in fections, fub-divifions, or in platoon.—The right flank being fuppofed to lead, on the word, HALT, FRONT, the platoon inftantly halts, and faces to its left; the CAUTION is then given, BY SECTIONS, SUBDIVISIONS, OR PLATOON, ON THE LEFT BACKWARD WHEEL, and at the word MARCH, the wheel ordered is made, in the manner directed in Sect. 48.

But in fituations where it may have been neceffary to order an extenfion of files, (fuch as will fometimes occur in marching through the ftreets of a town) a body thus moving, in order to avoid incorrect diftances

tances between the divisions, may form to the front in the following manner, either by platoon, sub-divisions, or sections.—On the word, TO THE FRONT FORM PLATOON; the front rank man of the leading file alone halts, and is instantly covered by his center and rear rank men: every other file of the platoon makes a half face to the left, and successively moving up, dresses on the right file; when the commander of the platoon sees it is properly dressed, he gives the word, *Eyes left*, and places himself on the pivot flank. *Front form platoon.*

Eyes left.

Should the order have been, TO THE FRONT FORM SUB-DIVISIONS (OR SECTIONS), the leading sub-division, or section, will proceed in the manner already detailed for the platoon; the succeeding sub-divisions, or sections, will each continue moving on, until its front file arrives at the proper forming distance from the division in its front, when it will receive from its commander the word, *To the front form*, and will instantly form up by files, in the manner already described. *Front form sub-divisions.*

Front form.

S. 61. *Forming from File to either Flank.*

The platoon marching in file (suppose from the right) has only to halt, and front, to be formed to the left flank.

To form to the right, it will receive the word, *To the right form*; the front rank man of the leading file, instantly turns to his right, and halts; his center and rear rank men at the same time move round and cover him.—All the other files of the platoon make a half turn to their left, and move round successively, in a line with the right hand file; *Right form.*

the

{ the center and rear rank men of each file, keeping clofed well up to their file leaders.

S. 62. *To form to either Flank, from Open Colum of Sub-Divifions.*

HALT.
LEFT WHEEL
INTO LINE.
MARCH.

The platoon marching in the ordinary time in open column of fub-divifions, to form to its left, receives the words, HALT, LEFT WHEEL INTO LINE.—MARCH, &c. and proceed as has already been fhewn in Section 50.

RIGHT FORM PLA-
TOON.

Halt, right wheel.
Halt, right drefs.

Left oblique.

Forward.

Halt, right wheel.
Halt, drefs up.

To form the platoon to its right flank, the inftructor of the drill gives the cautionary word of command, TO THE RIGHT FORM THE PLATOON; on which the commanders of the feveral divifions fhift to the right flank, and the commander of the leading fub-divifion, inftantly gives the word to his divifion, *Halt, Right wheel*; and when it has wheeled fquare, he orders, *Halt, right drefs*; goes to the right flank of his divifion, and dreffes it on the intended line of formation.—The commander of the other fub-divifion, on the leading one being ordered to wheel, gives the word, *To the left oblique*, and gradually inclines, fo as to be able to march clear of the rear rank of the divifion forming; this being effected, the word, *Forward* will be given to the divifion, and it will move on in the rear of the one formed.—When the fecond fub-divifion is arrived at the left flank of the firft, its commander gives the word, *Right wheel*, then, *Halt, drefs up*; on which the divifion moves up into the line, with the one formed; and its commander, from the left of his firft divifion, dreffes his own on the given flank point as quickly and as accurately as poffible, and refumes his proper platoon place.

S. 63.

S. 62. *The Platoon moving to the Front, to gain Ground to a Flank, by a March in Echellon, by Sections.*

In the drill of the platoon, when the soldier is compleatly formed, he may be Fig. 3. taught to march in echellon, by sections. This is a very useful movement for a battalion, or larger body moving in line, that is required to gain ground to a flank, and may be substituted instead of the oblique march.—It will be performed in the following manner.

The platoon marching to the front in the ordinary time, receives the word, BY SECTIONS TO THE RIGHT; the right hand men of the front rank of each section, turning in a small degree to their right, mark the time or three paces, during which the sections are wheeling in ordinary time on their pivot men; at the fourth pace, and at the word, *Forward*, the whole move on direct to the front that each section has now acquired, and the commander of each section, having taken post on the right of his division, the platoon continues its march in echellon.

SECTIONS, RIGHT.

FORWARD.

On the word, FORM PLATOON, the pivot men mark the time for three paces, turning back in a small degree to their left, the original front, and the sections instantly wheel backward into line; at the fourth pace the whole move forward. When the platoon is in two ranks only, two paces instead of three will be sufficient to mark time, and to step off at the third, instead of the fourth pace.

FORM PLATOON.

FORWARD.

S. 64.

S. 64. *From three Ranks forming in two Ranks.*

Form two deep.
Left face.
Quick march.

Fig. 9.

Halt, front.
Dress up.

{ The platoon halted, is ordered, FORM TWO DEEP; the rear rank men of the left sub-division instantly step back one pace; on the word LEFT FACE, the rear rank of both sub-divisions face; the word QUICK MARCH is then given, on which the men of the rear rank of the left sub-division step short, until those of the right get up to them; they then move on with them in file; as their rear is clearing the left flank of the platoon, the commander (who has shifted to this flank during the movement) gives the words *Halt, front, dress up,* he instantly dresses them on the standing part of his platoon, and resumes his post on the right.—One third, or one more sub-division, is thus added to the front of the company.

If a battalion is standing in open column, it may thus encrease the front of its companies, before it forms in line;—But if it is already in line and is thus to encrease its front, its companies must take sufficient intervals from each other, before their respective rear ranks can come up.——If a battalion in line is posted, and without deranging its front is to lengthen out a flank by the aid of its rear rank; it would order that rank to wheel backwards by sub-divisions: The last sub-division of each company would close up to its first one: All the sub-divisions (on the head one) would move forward to open column: An officer would be named to command those of each two companies: The open column would move on, and wheel into line on the flank of the battalion.—In this manner also would a line of several battalions lengthen itself out by the rear ranks of each.

§. 65.

S. 65. *From two Ranks forming into three Ranks.*

The platoon being halted and told off into three sections, it receives the word FORM THREE DEEP; on which the third section instantly steps back one pace; the word RIGHT FACE is then given, and the man on the right of its front rank, on facing, disengages a little to his right; on the word QUICK MARCH, the front rank men of the third section step off, those of the other rank mark the time till they have past, and then follow.—When the leading man has got to the right of the platoon, the commander gives the word *Halt, front,* on which each man halts, faces to his left, and instantly covers his proper file leader.

Fig. 9.

FORM THREE DEEP.

RIGHT FACE.

QUICK MARCH.

Halt, Front.

A rear rank which has lengthened out, and formed on the flank of its battalion, would return to its place: By wheeling back into open column of sub-divisions; marching till each arrived at its flank point; the leading rank of each would wheel up and cover; and the second rank would move behind it, and also wheel up.

S. 66. 67. 68. 69.

In pursuance of the foregoing instructions, and on the principles they contain, every company of a battalion must be frequently exercised by its own officers, each superintending a rank, or an allotted part of the whole.—And on a space of 70 or 80 yards square, can every circumstance be practised that is necessary to qualify it for the operations of the battalion.—That space being pointed out by under officers, or other marks, as directed at the latter end of the first part, the company will, both at open and close files, without arms, and with arms.

Exercise of Company.

By

By Ranks.

1. March in single file, by successive ranks, along the 4 sides of the square.—The same, by two's.

2. March, and wheel, by ranks of fours;—File off singly and double up, preserving proper distances, and not quickening on the wheel.

3. March, and wheel, by sub-divisions of ranks.

4. March, and wheel, by whole ranks.

5. March to front, and to rear; ranks at 10 paces asunder.

6. March the company in a single rank, to front, and to rear, by a flank, and by the center.

7. Oblique by ranks.

8. Open, and close files, and intervals, by the side step.

9. March in file, to either flank.

10. Ranks successively advance 6 or 8 paces; halt, and dress.—Ranks successively fall back 6, or 8 paces; halt, and dress.

11. Advance, or retire 2 or 3 flank men; the ranks dress to them.

12. Open, and close ranks.

At Close Ranks, and Files.

13. March, and wheel in all directions, by sub-divisions, and by company.—Shorten step, and lengthen it, the march to be made both in ordinary and quick time.—The wheels to be made in wheeling time.

14. Advance, and retire, 2 or 3 flank files, and dress to them.

15. Open, and close to the flank, by the side-step.

16. Change front by the counter march by files.

17. March to the flanks, close and without opening out.—Form to the front, or to either flank.

18. March oblique.

19. Sub-divisions double on the march, and again form up, by obliquing.

20. Wheel backwards by sub-divisions.—March along the line, to prolong it:—form to the flank, by wheeling up; or to the front by obliquing.

21. File from the flank of company to the rear, as in the passage of lines:—*Halt, front;*—Close into pivot file:—Wheel up, as in forming in line.

22. From 3 deep, form 2 deep.

23. From 2 deep, form 3 deep.

24. Exercise of the firelock, manual, and platoon, by ranks, and company.

25. Firings by files, sub-divisions, and company.

The necessary pauses, and formations, betwixt these movements, in order to connect them, must of course be made.—They may be practised in whatever succession shall at the time be found proper.—The greatest precision must be required, and observed, in their execution, according to the rules already laid down.

Every officer must be instructed in each individual circumstance required of a recruit, or a soldier; also in the exercise of the sword; and accustomed to give words of command, with that energy, and precision, which is so essential.—— Every officer, on first joining a regiment, is to be examined by the commanding officer; and, if he is found imperfect in the knowledge of the movements required from a soldier, he must be ordered to be exercised that he may learn their just execution. Till he is master of those points, and capable of instructing the men under his command, he is not to be permitted to take the command of a platoon in the battalion.

Squads of officers must be formed, and exercised by a field officer; they must be marched in all directions, to the front, oblique, and to the flank; they must be marched in line, at platoon distance, and preserve their dressing and line from an advanced center: they must be placed in file at platoon distance, and marched as in open column; they must change direction, as in file, and cover anew in column. In these, and other similar movements, the pace and the distances are the great objects to be maintained.—From the number of files in division, they must learn accurately to judge the ground necessary for each, and to extend that knowledge to the front of greater bodies. They must acquire the habit of readily ascertaining, by the eye, perpendiculars of march, and the squareness of the wheel.

An officer must not only know the post, which he should occupy, in all changes of situation, the commands which he should give, and the general intention of the

required

required movement; but he should be master of the principles, on which each is made; and of the faults that may be committed, in order to avoid them himself, and to instruct others.—These principles are in themselves so simple, that moderate reflection, habit, and attention, will soon show them to the eye, and fix them in the mind; and individuals, from time to time, when qualified, must be ordered to exercise the battalion, or its parts.

The complete instruction of an officer enlarges with his situation, and at last takes in the whole circle of military science:—From the variety of knowledge required of him, his exertion must be unremitting, to qualify himself for the progressive situations at which he arrives.

Besides the instruction peculiar to the under officers, they should be exercised in the same manner as the officers are, as they are frequently called on to replace them:—The necessity also of order, steadiness, silence, and of executing every thing deliberately, and without hurry, should be strongly inculcated in the infantry soldier.

D. D.

End of Part Second.

PART the THIRD.

OF THE BATTALION.

A perfect Uniformity in the Formation and Arrangement of all Companies and Battalions, is indifpenfible for the Execution of juft and combined Movements.

FORMATION OF THE COMPANY.

THE Company is always to be fized from flanks to center.

The Company is formed three deep.

The Files lightly touch when firelocks are fhouldered and carried, but without crouding; and each man will occupy a fpace of about twenty-two inches.

Close order is the chief and primary order, in which the battalion and its parts at all times assemble and form.—Open order is only regarded as an exception from it, and occasionally used in situations of parade and shew.—In close order; the officers are in the ranks, and the rear ranks are closed up within one pace. In open order; the officers are advanced three paces, and the ranks are two paces distant from each other.

Each company is a platoon.—Each company forms two subdivisions, and also four sections. But as sections should never be less than five files, it will happen, when the companies are weak, that they can only (for the purposes of march) form three sections, or even two sections.

When the company is singly formed; the captain is on the right, and the ensign on the left, of the front rank, each covered by a serjeant in the rear rank. The lieutenant is in the rear, as also the drummer and pioneer in a fourth rank, at three paces distance.

The left of the front rank of each subdivision is marked by a corporal. The right of the left subdivision may be marked by the other corporal.

When necessary, the places of absent officers may be supplied by serjeants, those of serjeants by corporals, and those of corporals by intelligent men.

When the company is to join others, and the battalion, or part of it, to be formed; the ensign and his covering serjeant quit the flank, and fall into the fourth rank, until otherwise placed.

§. 70. *When the Company is to take Open Order from Close Order.* — Commands.

At this command, the flank men on the right and left of the rear ranks, step back to mark the ground on which each rank respectively is to halt, and dress at open distance; they face to the right, and stand covered; every other individual remains ready to move. — *Rear ranks take open order.*

At this command, the rear rank dressers front, and the rear ranks fall back one and two paces each dressing by the right, the instant it arrives on its ground: The officers move out in front three paces, and divide their ground: One serjeant is on each flank of the front rank: The pioneer remains behind the center of the rear rank: The drummer places himself on the right of the right serjeant. — *March.*

---

§. 71. *When the Company is to take Close Order from Open Order.*

The officers, serjeants, drummer, face to the right. — *Rear ranks take close order.*

The ranks close within one pace, marching one and two paces, and then halting. — *March.*

{ The officers move round the flanks of the company to their respective posts: The serjeants and drummers fall back, and each individual resumes his place, as in the original close order.

The above regards the company when single; but when united in the battalion, other posts are allotted to the drummer and pioneer.

FORMATION of the BATTALION.

Strength of the battalion.

The battalion is ten companies, { 1 Grenadier, 8 Battalion, 1 Light,

Each company consists at present of { 3 Officers, 2 Serjeants, 3 Corporals, 1 Drummer, 30 Private.

Formation of the battalion.

When the companies join and the battalion is formed, there is to be no interval between any of them, grenadier, light company, or other; but every part of the front of the battalion should be equally strong.

Each

Each company which makes a part of the same line, and is to act in it, must be formed and arranged in the same manner.

The companies will draw up as follows from right to left :—grenadiers;—1st captain and major;—4th and 5th captain;—3d and 6th captain;—2d captain and lieutenant colonel;—light company.—— The colonel's company takes place according to the rank of its captain: The four eldest captains are on the right of the grand divisions: officers commanding companies or platoons, are all on the right of the front rank of their respective ones. *Position of the companies in battalion.*

The eight battalion companies will compose four grand divisions;— *Divisions.* eight companies or platoons,—sixteen subdivisions,—thirty-two sections, when sufficiently strong to be so divided; otherwise twenty-four, for the purposes of march.—The battalion is also divided into right and left wings.—When the battalion is on a war establishment, each company will be divided into two platoons.—When the ten companies are with the battalion, they may then, for the purposes of firing or deploying, be divided into five grand divisions from right to left.

The battalion companies will be numbered from the right to the left, 1. 2. 3. 4. 5. 6. 7. 8.—The subdivisions will be numbered 1. 2. of each; —the sections will be numbered 1. 2. 3. 4. of each;—the files of companies will also be numbered 1. 2. 3. 4. &c.—The grenadier and light companies will be numbered separately in the same manner, and with the addition of those distinctions.—These several appellations will be preserved, whether faced to front or rear.

The

Companies equalized.

The companies must be equalized in point of numbers, at all times when the battalion is formed for field movement, and could the battalions of a line also be equalized, the greatest advantages would arise; but though from the different strengths of battalions this cannot take place, yet the first requisite always must, and is indispensible.

Formation of the battalion at close order.

Ranks are at the distance of one pace, except the fourth or supernumerary rank, which has three paces.

All the field officers and the adjutant are mounted.

The commanding officer is the only officer advanced in front, for the general purpose of exercise when the battalion is single; but in the march in line, and in the firings, he is in the rear of the colours.

The lieutenant colonel is behind the colours, six paces from the rear rank.

The major and adjutant are six paces in the rear of the third and sixth companies.

One officer is on the right of the front rank of each company or platoon, and one on the left of the battalion; all these are covered in the rear rank by their respective serjeants; and the remaining officers and serjeants are in a fourth rank behind their companies.—It is to be observed, that there are no coverers in the center rank to the officers or colours.

The

The colours are placed between the fourth and fifth battalion companies, both in the front rank, and each covered by a non-commissioned officer, or steady man in the rear rank.—One serjeant is in the front rank betwixt the colours; he is covered by a second serjeant in the rear rank, and by a third in the supernumerary rank.—The sole business of these three serjeants is, when the battalion moves in line, to advance and direct the march as hereafter mentioned.——The place of the first of those serjeants, when they do move out, is preserved by a named officer or serjeant, who moves up from the supernumerary rank for that purpose.

The fourth rank is at three paces distance when halted, or marching in line.—When marching in column, it must close up to the distance of the other ranks.——The essential use of the fourth rank, is to keep the others closed up to the front during the attack, and to prevent any break beginning in the rear; on this important service, too many officers and non-commissioned officers cannot be employed.

Use of the fourth or supernumerary rank.

The pioneers are assembled behind the center, formed two deep, and nine paces from the third rank.

The drummers of the eight battalion companies are assembled in two divisions, six paces behind the third rank of their second and seventh companies.——The grenadier and light company drummers and fifers, are six paces behind their respective companies.

The music are three paces behind the pioneers in a single rank, and at all times, as well as the drummers and pioneers, are formed at loose files only, occupying no more space than is necessary.

The

The staff of chaplain, surgeon, quarter-master and surgeon's mate, are three paces behind the music.

Officers.

In general, officers remain posted with their proper companies; but commanding officers will occasionally make such changes as they may find necessary.

Replacing serjeants.

Whenever the officers move out of the front rank, in parade, marching in column, wheeling into line, or otherwise, their places are taken by their serjeant coverers, and preserved until the officers again resume them.

When the line is halted, and especially during the firings when engaged; the serjeant coverers fall back into the fourth rank, and observe their platoons.

Commands.

Rear ranks take Open Order.

S. 72. *When the Battalion takes open Order.*

At this command—the flank men on the right of the rear ranks of each company step briskly back to mark the ground on which each rank respectively is to halt. They *face* to the right, and cover as pivots, being regulated and dressed by the adjutant or serjeant major on the right.—— Every other individual remains ready to move.

At

At this command—the flank dreffers face to the front, and the whole move as follows:

The rear ranks fall back one and two paces, each dreffing by the right the inftant it arrives on the ground.

The officers in the front rank, as alfo the colours, move out three paces—thofe in the rear, together with the mufic, move through the intervals left open by the front rank officers, and divide themfelves, viz. the captains covering the fecond file from the right, the lieutenants the fecond file from the left; and the enfigns oppofite the center of their refpective companies.

The mufic form between the colours, and the front rank. *March*,

The ferjeant coverers move up to the front rank, to preferve the intervals left by the officers.

The pioneers fall back to fix paces diftance behind the center of the rear rank.

The drummers take the fame diftance behind their divifions.

The major moves to the right of the line of officers.—— The adjutant to the left of the front rank.

The ftaff place themfelves on the right of the front rank of the grenadiers, viz. chaplain, furgeon, quarter-mafter, mate.

The lieutenant-colonel, and the colonel (difmounted), advance before the colours, two and four paces.

The whole being arrived at their several posts—Halt—Dress to the Right—and the battalion remains formed in parade, in the order in which they would receive a superior officer.

When the battalion is reviewed singly, then in order to make more show—the division of drummers may be moved up, and formed two deep on each flank of the line—the pioneers may form two deep on the right of the drummers of the right—and the staff may form on the right of the whole.

S. 73. *When the Battalion resumes Close Order.*

Rear ranks take close order.

The lieutenant-colonel, officers, colours, staff, music, face to the right.

The drummers and pioneers (if on the flanks) face to the center.

The serjeants (if in the front rank) face to the right.

March.

The rear ranks close within one pace, moving up one and two paces, and then halting.

The music marches through the center interval.

The serjeants, drummers, pioneers, &c. &c. resume their places, each as in the original formation of the battalion in close order.

The officers move through and into their respective intervals, and each individual arrives at, and places himself properly at his post in close order.

On particular occasions, and when necessary, officers commanding platoons, who in line are on the right of their platoons, shift to the left to conduct the heads of files, or the pivot flanks of their divisions in column or echellon. Posting of officers.

When the battalion wheels by companies or subdivisions to either flank into column; both colours and the file of directing serjeants always wheel to the proper front, and place themselves behind the third file from the new pivot. Colours.

There is no separate colour reserve; the pioneers, music, &c. sufficiently strengthen the center; but in the firings the two files on each side of the colours may be ordered to reserve their fire. Colour reserve.

The constant order of the light company when formed in line, and united with the battalion, is at the same close files as the battalion.—— Their extended order is an occasional exception. Light company.

When the light company is detached, and the grenadier company remains, it will be undivided on one flank of its battalion, whenever there are several battalions in line: but when the battalion is single, it is permitted to be occasionally divided on each flank. Grenadiers.

When the grenadier or light companies are detached, and make no part of the line, they may be formed two deep, if it is found proper.

With a very few obvious alterations, thefe general rules take place when a company or battalion is permitted or ordered to form in two ranks only---and which on the prefent low eftablifhment of our battalions, may often be done for the purpofes of exercife and movement on a more confiderable front: it is alfo evident that they generally apply whether the companies are ftrong or weak, and whether a greater or leffer number of them compofe the battalion.

GENERAL CIRCUMSTANCES PREPARATORY TO THE MOVEMENTS OF THE BATTALION.

S. 74. *Commands.*

All words of command, and particularly the words HALT or MARCH muft be given, fhort, quick and loud, fo as to be caught and repeated from right to left of a line, or from front to rear of a column, in the fhorteft time poffible.

All alterations, in carrying ARMS; change of PACE; WHEELING; FACING; OBLIQUING; HALTING; MARCHING; and in general every operation of the battalion whether in line or column, which ought to be executed by the whole battalion at the fame inftant, are made by each in confequence of one word from the commander of each: but there are alfo many occafions in column, and in forming line, where the leaders of divifions repeat, or give the words of *March, Wheel, Halt, Front, Drefs,* &c. to their feveral divifions, as is neceffary.

Every

Every officer muſt be accuſtomed to give his words of command, even to the ſmalleſt bodies in the full extent of his voice, and in a ſharp tone: By ſuch bodies he muſt not only be heard, but by the leaders of others who are dependant on his motions.———The juſtneſs of execution, and the confidence of the ſoldier can only be in proportion to the firm, decided, and proper manner in which every officer of every rank, gives his orders.———An officer who cannot thoroughly diſcipline and exerciſe the body entruſted to his command, is not fit in time of ſervice to lead it to the enemy; he can not be cool, and collected in the time of danger; he cannot profit of circumſtances from an inability to direct others; the fate of many depends on his ill or well acquitting himſelf of this duty,—It is not ſufficient to advance with bravery; it is requiſite to have that degree of intelligence, which ſhould diſtinguiſh every officer according to his ſtation: nor will ſoldiers ever act with ſpirit and animation, when they have no reliance on the capacity of thoſe who do conduct them.

In the midſt of ſurrounding noiſes, the eye and the ear of the ſoldier ſhould be attentive only to his own immediate officer; the loudneſs of whoſe commands inſtead of creating confuſion and unſteadineſs, reconcile to the hurry of action.

On all occaſions when words of command are not heard, if the directing body has made a change of ſituation, the reſt of the body will conform to it, as ſoon as the intelligence of the officer has pointed out, what is meant to be done; and the eye will often ſhow the propriety and moment of movement, when the ear has not received the explanatory command.

The

The field officers and adjutant of the battalion are at all times mounted.——In order the more readily to give ground in movements, speedily to correct mistakes, to circulate orders, to dress pivots, when they ought to cover in column in a straight line, and especially to take care when the column halts, that they are most speedily adjusted before wheeling up into line.——These operations no dismounted officer can effectually perform, nor in that situation can he see the faults, or give the aids which his duty requires.

S. 75. *Distance of Files.*

Except in the instruction of recruits and squads, on some occasions of regimental parade or inspection, and in the peculiar exercise of the light company, open files are not to be used, and at all times the battalion, or its most minute parts are to form, move, and act at CLOSE FILES, so that each soldier when in his true position under arms, shouldered, and in rank, must just feel with his elbow the touch of his neighbour with whom he dresses, nor in any situation of movement in front, must he ever relinquish such touch, which becomes in action, the principal direction for the preservation of his order, and each file as connected with its two neighbouring ones, must consider itself a compleat body so arranged for the purpose of attack, or effectual defence.

It cannot be doubted, when a battalion arrives at its object of attack at CLOSE FILES, that both its impulse and quantity of fire in the same extent of front is greater, than when the files are more open; and

should

should crowding be apprehended, it is at all times more eligible to have a division obliged to fall out of the line and double, than to have openings in it, where the enemy must certainly penetrate.

The perfect and correct march of a battalion or line formed at OPEN FILES seems hardly attainable, because its principal guidance, the touch of the files does not exist; each man is necessarily employed to preserve a required distance from his neighbour, he is obliged for that purpose to turn his head, this distorts his body, and gives him a direction contrary to the perpendicular line he should march on, a constant opening and closing takes place, and the whole move loose and unconnected.— If this must necessarily happen in a single battalion, the influence on a line may be easily imagined, and also the condition in which it will arrive near an enemy; who if he is formed at CLOSE FILES, if his dressing and line are chiefly determined by the touch; if the eyes alone are glanced towards the center of battalions; if the figure of each individual is full to the front, if the whole move square along their just lines without crouding at an uniform and cadenced pace, which habit alone, unchecked by false and adventitious aids has given: He at every instant of movement or attack will be firm, united and animated with that sense of his own superiority which perfect order, and due consistence will always give.

S. 76. *Distances of Ranks.*

There are two distances of ranks, *Open* and *Close*.—When open they are two paces asunder.—When close they are one pace. When the body is halted and to fire, they are still closer locked up.

Close ranks or order is the constant and habitual order at which troops are at all times formed and move.—*Open* ranks or order, is only an occasional exception, made in situations of parade.

The distances of files and ranks relate to the trained soldier, but in the course of his tuition he must be much exercised at open files and ranks to acquire independence and the command of his limbs and body.

S. 77. *Depth of Formation.*

The fundamental order of the infantry, in which they should always form and act, and for which all their various operations and movements are calculated, is in *three ranks*: The formation in *two ranks*, is to be regarded as an occasional exception that may be made from it, where an extended and covered front is to be occupied, or where an irregular enemy who deals only in fire is to be opposed.——But from the present low establishment of our battalions, they are during this time of peace permitted, in order to give the more extent of front in their operations, to continue to form and use it, in many of their movements and firings, at the same time not omitting frequently to practise them in three ranks.

The formation in two ranks, and at open files, is calculated only for light troops in the attack and pursuit of a timid enemy, but not for making an impression on an opposite regular line, which vigorously assails, or resists.——No general could manage a considerable army if formed and extended in this manner.——The great science and object of movement being to act with superiority on chosen points; it is never the intention of an able commander to have all his men at the same time in action; he means by skill and manœuvre to attack a partial part, and to bring the many to act against the few; this can not be accomplished by any body at *open* files, and *two* deep.—A line formed in this manner would never be brought to make or to stand an attack with bayonets, nor could it have any prospect of resisting the charge of a determined cavalry.—In no service is the fire and consistency of the third rank given up; it serves to fill up the vacancies made in the others in action, without it the battalion would soon be in a single rank.

S. 78. *Musick and Drums.*

The use of MUSICK or DRUMS to regulate the march is absolutely forbid, as incompatible with the just and combined movements of any considerable body, and giving a false aid to the very smallest.—They never persevere in the ordered time or in any other, are constantly changing measure, create noise, derange the equality of step, and counteract the very end they are supposed to promote. The ordered and cadenced MARCH can be acquired and preserved from the eye and habit alone, and troops must by great practice be so steadied as to be able to maintain it, even though drums, musick, or other circumstances, should be offering a different marked time.—On occasions of parade and show, and when troops are halted, they are properly used, and when circumstances do not forbid it, may be sometimes permitted as infpiriting in column of march, where unity of ftep is not fo critically required.—But in all movements of manœuvre whatever, and as at any time directing the cadence of the ftep, or in the inftruction of the recruit, officer or battalion, they muft not be heard.

S. 79. *The March.*

General intention.

1. All military movements are intended to be made with the greateft quicknefs, that is confiftent with order, regularity, and without hurry or fatigue to the troops.——The uniformity of pofition, cadence, and length of ftep produce that equality and freedom of march, on which every thing depends, and to which the foldier muft be carefully trained, nor fuffered to join the battalion until he is thoroughly perfected in this moft effential duty.

Degrees of march.

2. The different degrees of march have been already detailed in the firft, and fecond parts, and to thefe muft the foldier be trained and accuftomed without drum, or mufic, and by habit alone taught to acquire the given times, and length.—To the equal and unvaried cadence and length of ftep thus attained, can troops alone truft for the prefervation of their line in advancing upon an enemy, when duft, the fmoke of artillery, rain, fog, and many other local circumftances, make it impoffible to depend on diftant points, the uncertain time given by timid muficians, or any other adventitious help.

3. A company

3. A company or division may occasionally run, a battalion may sometimes march quick, but the hurrying of a large column, or of a body moving in front, will certainly produce confusion and disorder. It is never to be risqued when an enemy is in presence and to be encountered; though it may sometimes be necessary where a post or situation is to be seized.

---

4. The use of the side or closing step has been already mentioned S. 43.——If more than one platoon is to close, at the words, &c. Close—March—Halt, the closing body proceeds as directed S. 13. 43. If the body which is to close is truly formed, and has no false openings in it which are to be corrected, but that the whole is meant to be shifted to a named flank; the word from the commanding officer puts in march, and halts the whole.—But if the intention is to correct improper intervals between platoons or files: The word from the commanding officer puts the whole that are to close in march towards the ordered hand; and each platoon officer separately and successively gives his word *Halt* at the proper instant that his platoon has closed to that hand; this he is the better enabled to do from being himself out of the rank, and facing his platoon.

Side step.

---

5. All halts are made to the point, to which the troops while in march are looking; by bringing up the rear foot to the advanced one, so as to finish the step which is taking, when the command is given; and after which no dressing or movement whatever is to be made, until a separate order directs it.

Halts.

Oblique march.

Fig. 1.

6. The oblique march enables a body to preserve its parallel direction, and at the same time to gain ground to the flank, as well as to the front without filing or opening out.—It is particularly necessary for the battalion in line, when intervals are to be corrected, and in the forming up, and doubling of its divisions.—With a body of any extent it is a very nice operation to execute.—Each battalion in line obliques without turning eyes from its own center.——One degree of obliquing only (under the angle of about 25°.) is to be required from an extended front of troops, and even in that it is exceeding difficult to preserve them: but the smaller divisions of the battalion will often be obliged in forming up, or in doubling, especially when in movement, to oblique more or less sharply, according to circumstances.——S. 8. 35. 55.

S. 80. *Wheeling.*

Wheeling in general.

1. A single rank or division might at all times wheel to a *halted flank*, without alteration of the time at which it is then marching; by the outward wheeling man preserving the usual length of step, and the others properly shortening theirs to remain dressed with him: The same might take place in column, whose divisions were equal, and when the wheel is under 1-6th of the circle.—But when it exceeds that portion, it becomes necessary in order to clear the ground, prevent false distances, and a lengthening out; that the divisions successively make their wheels to their *halted flank* at a pace considerably quicker, than what the body of the column is moving at.

2. Wheels

2. Wheels of divisions of a battalion or line, are made on a halted pivot, or on a moveable pivot.—When on a HALTED pivot they are made from line into column, or from column into line; and also generally by the column of manœuvre or march in movement, when the front of it is considerable, and when the wheel by which its direction is to be changed approaches to, or exceeds the quarter circle.—When on a MOVEABLE pivot, they are only used and occasionally ordered in the column of march, when its front is small, and that its path is winding, and changeable; in that case both flanks are moveable and describe concentric circles round a point which is removed a few paces, from what would otherwise be the standing flank. *Wheels made on halted or moveable pivots.*

Fig. 4. B.

3. The various circumstances attending the wheels on the HALTED pivot have been detailed in the first, and second parts.——Altho' the pause made after the *Halt,—Dress*, gives time in large fronts, for exact dressing; yet in small ones where that pause is short, there is no time for such exactness, the attention to, and preservation of the true distance being then the material object.—Whenever the wheel made is less than the quarter circle the pause after the wheel will be considerable; should the wheel be greater than the quarter circle it must be accelerated, otherwise more than one division will be arrived and arrested at the pivot point.——Should a column be marching in quick time, it is evident that its wheels must be in proportion quickened to disengage in time the pivot point, for each successive division. *On a halted pivot.*

Fig. 4. B.

4. When the column of sections, subdivisions, or companies, is obliged frequently to change its direction of march, and that it is permitted *On a moveable pivot.*

mitted to do it on the MOVEABLE pivot (S. 21. 52.) inftead of a halted pivot.—If the pivot leader defcribes the fmaller circle of the wheel, he leaves the point on which he marched, and where the old, and new directions interfect, clofe to his own hand wheeled to.—When he defcribes the greater circle, he leaves fuch point wide from his own wheeling hand.—In both cafes the more confiderable the fweep he makes from the old to the new direction the eafier, and more gradually can the other flank conform, and therefore when this mode is made ufe of, the column is fuppofed to have fufficient room on its flanks to allow of the neceffary operations; for if both flanks cannot be kept in progreffive movement during the change of direction, the wheels cannot be thus made, but muft be executed quick, and on *fixed* pivots, otherwife the ground would not be clear for the fucceeding divifions, and they would ftop each other, and interfere.—In this manner will the column on a fmall front follow the windings of a route, be conducted through an open wood, or trace out the irregular edge of a height, which it is to occupy; and indeed on all common occafions of route marching, where perfect correctnefs is not required, it will thus change into new directions.

Fig 4. B.

Wheel of divifions backwards.

5. Wheels of divifions may be made either forward, or backward.—In general, (and always in progreffive movement) they are made forward, but particular occafions require that they fhould be made BACKWARD, on the pivot flank.——In this manner may the line wheel into open column of platoons, fubdivifions, or fections: the flank fides of the fquare, or oblong may thus wheel into column, when the body is to be put in march: the line already formed may be thus prolonged when neceffary to either flank, as the pivots are thus preferved: it is

alfo

also advantageously used in marching off parades, where guards are of different strengths, and is often essentially necessary in narrow grounds. —By this means although divisions should be unequal, either in the same battalion, or in a line, yet all their pivot flanks will after the wheel remain truly dressed; of course the distances will be just, the line of marching accurately preserved, and each division by afterwards wheeling up will exactly occupy the identical ground it quitted.—— Whereas in wheeling FORWARD from line into open column, even if the divisions are of equal strength, the pivots and distances after the wheel will not be true, because the different sizes of men, and the least over or under wheel of any one division will derange them, which in practise will infallibly happen.—But if the divisions are of unequal strength, independant of the pivots necessarily not covering, the distances which the column marches off at, must be all changed during the march; otherwise when the column is to wheel up, and form, strong divisions would have to wheel into the space, which the weaker ones had left, and *vice versa*; the consequence and confusion thence arising, is obvious.

Fig. 21. B.

Fig. 21. A.

6. To prevent therefore such inconvenience it must be regarded as a rule almost general—That all wheels by companies or smaller divisions from battalion or line (when halted) into open column should be made BACKWARD, and all wheels from open column into line FORWARD: The only necessary exceptions seem to be in some cases in narrow grounds where there is not room for such wheels.—If the division does not exceed 16 or 18 file, it may readily wheel back without facing about; but if the division is stronger and the ground uneven, it must *Face about*—*Wheel*—and then *Halt, front*.

General rule.

7. In

When wheels of divisions are made backwards, or forwards.

7. In wheeling BACKWARD from line into column; when the right is to be in front, the wheel is made ON the left; and when the left is to be in front, the wheel is made ON the right.—In wheeling FORWARD the standing flank man faces outward from his division: In wheeling BACKWARD, he faces inward to his division.—In wheeling FORWARD the proper pivot flank of the column is the wheeling one: In wheeling BACKWARD, the pivot flank is the standing one, and remaining fixed the divisions however unequal will always cover on that hand, which will not be the case if the wheel is made forward.—In wheeling FORWARD, the command is TO THE RIGHT, (or) TO THE LEFT, WHEEL: In wheeling BACKWARD, the command is ON THE RIGHT, (or) ON THE LEFT, BACKWARD WHEEL.

Circumstances in wheeling.

8. As the circumference of the quarter circle which a division describes in its wheel, is one half more, (nearly) than its front; it is necessary that in open column, it should, in the time that it takes to march over a space equal to the extent of its front, not only compleat the wheel of the quarter circle, but be enabled to move on at its just distance from its preceding division, and not to stop that which succeeds it. The wheel must therefore be quickened, or the step lengthened, (or part of both applied) in proportion to the general march.

Number of files in a division, each occupying 22 inches.	5.	10.	12.	14.	15.	16.	18.	20.	30.	40.	50.	100.
Front of divisions in ordinary paces of 30 inches.	P. In. 3. 20	7. 10	8. 24	10. 8	11.	11. 22	13. 6	14. 20	22.	29. 10	36. 20	70. 10

9. A division

9. A division consisting of 10 files, and each occupying 22 inches, will at paces of 30 inches, take 7 paces, 10 inches for its front.—Now 75 steps in a minute being the ordinary time, and 120 the wheeling time, $75 : 120 :: 7\frac{1}{3} : 11\frac{2}{3}$ nearly the number of wheeling paces of 30 inches each, which the wheeling man can take while the following division is making its $7\frac{1}{3}$ ordinary paces in front, and 11 of which exactly compleats the quarter circle: but if each of these 11 paces is lengthened with 3 inches, then the wheel will be compleated in 10 steps, and a pause of one pace and 2-3ds of a pace, or 5-6ths of a second of time will be reserved for the *Halt, Dress,* and *March* of the division after it has at 10 long paces of 33 inches compleated the wheel.—This pause will encrease or diminish according to the greater or lesser extent of the wheeling body, and in the above proportions of time and step, it is 1-7th of the time employed by such body in wheeling the quarter circle.——This allowance which is barely sufficient in a division of 10 files, and which cannot well be encreased, either by length of step, or quickness of time shows how pointed and quick the commands must be, not to occasion a loss of ground to each successive division at the points of wheeling.

10. It appears that the front of any division or body, is in ordinary paces of 30 inches, nearly 3-4ths of the number of files of which it is composed.——That the circumference of the quarter circle which it describes is in wheeling paces of 33 inches, the same as the number of files of which it is composed,——That the number of files being once ascertained in each division, the officer commanding it must on all occasions recollect the number of paces that are equal to his front; also the number of wheeling paces which the flank man must take to compleat

Necessary recollections.

the quarter circle; also the spare time, which he has to regulate the *Halt, March* of his division after wheeling.

Wheeling paces required to describe				
	The 6th of the circle, or an angle of 60°.	are $\frac{2}{3}$	of the number of files of which the front consists.
	The 8th	45°.	$\frac{1}{2}$	
	The 16th	22°.$\frac{1}{2}$	$\frac{1}{4}$	
	The 32d	11.$\frac{1}{4}$	$\frac{1}{8}$	

11. The field officers and adjutants, must always recollect the number of paces the front of the battalion and its divisions occupy, in order to take up ground exactly in all formations.

S. 81.   *Movements.*

1. Every movement must be divided into its distinct parts, and each part executed by its explanatory and separate words of command.

2. Alterations of position in considerable bodies should begin from a previous halt; except giving a new direction to the heads of columns, or encreasing or diminishing their front which may be done while in motion.

3. The exercise of small bodies when within the command of one voice, appears more showy from the keeping such bodies constantly in motion, and by changing from one manœuvre to another while on

the

the march.—But such movements and the formations made from them must be on accidental points, and however brilliant in battalion practice, and review appearance, where the lesson of the day has been previously arranged, they can only be considered as occasional exceptions not applicable to large bodies where hurry must be avoided, and where concert, and relative position are indispensible.

4. As the principle of moving, forming, and dressing upon given, and determined points is just, all quick alterations of position of a considerable body, attempted while on the move, and not proceeding from a previous halt (however short) are false, and defective, the effects of which though not so apparent in a single battalion would be very obvious in a line or column of any extent.—A pause between each change of situation so essentially necessary to the movements of great bodies, should seldom be omitted in those of small ones; squareness of dressing, the exact perpendiculars of march, and the correct relative position of the whole are thereby ascertained.—Such alterations of situation made from the halt, may when necessary succeed each other quickly; and in many cases no unnecessary time need be taken up in scrupulous dressing, but every one may be instantly apprized of the following movement, which circumstances require.

S. 82. *Points of March.*

1. Every leader of a body which is to move directly forward in front, must take care to conduct it in a line truly perpendicular to that front.—

To march straight on one object only with certainty and without wavering, is not to be depended on; two objects therefore placed and preserved during the march in the same straight line are necessary for the purposes of correct movement, when the intent is truly to prolong a given line.

2. Two objects will therefore in general be prepared for the direction of any considerable body: But should a leader either in file or in front, have only one marked point of march ascertained to him, he will himself instantly look out for his small intermediate points, which are always to be found, which he will from time to time renew, and which are to preserve and determine the accuracy of his movement towards the more distant point. (V. S. 42.)

S. 83. *The Alignement.*

1. To march or form in the ALIGNEMENT, is to make troops march, or form in any part of the straight line which joins two given points.—On the justness and observance of this line, depends the accuracy of the most essential movements and formations, and therefore every relative help must be applied to ensure it.

2. In formations of defence the lines occupied may be curved, and following the advantages of the ground, but in those of attack, the lines must be straight, otherwise the troops in advancing must inevitably fall into confusion.

3. When troops are to form in a straight line, two necessary points in it must always be previously ascertained.—One the point of APPUI (A. a. a.) at which one flank of the body whether small or great, is to be placed, and the other the point of FORMATION or DRESSING (D.) on which the front of the body is directed. Fig. 12.

4. When battalions, or divisions of a battalion come up successively into line, the outward flank of the last formed and halted body, is always considered as the point of APPUI (a. a.) or support of the succeeding one, and in this manner is the general line prolonged from each successive point of Appui, towards the given distant point of formation (D).—The looking, and lining of the soldier in forming is always towards the point of Appui, and the correction of dressing is always from that point towards the opposite hand.—This great principle is to be observed, from the smallest body to the most considerable corps, and regulates the formation of the division, the battalion, and the line. Fig. 12.

S. 84. *Points of Formation.*

1. In the movements of a single battalion, and in the taking up of a new position, it may not seem material whether a flank is placed a few yards to the one hand or other, or whether the line formed on, is exactly directed on any certain point.—But when a battalion makes a part of a more considerable body, then all its positions being relative to other battalions and to given points, if its formations are not accurate

and

and juft, it will create general confufion and give falfe directions and diftances to thofe whofe fituation muft be determined by it.—The ne-
ceffity therefore of every fingle battalion being accuftomed to make its changes of pofition, and formations on determined points is apparent, and is an object which commanding officers muft always hold in view, and have their adjutants and others prepared and inftructed accordingly.

<small>Neceffity of formations made on given points.</small>

<small>Bafe line, and method of prolonging it.</small>

2. The line on which, troops in column move, or are fucceffively to form, is taken up to any extent by the prolongation of an original fhort and given BASE, eftablifhed where they firft begin to enter, or form on that line, the direction of which is determined by the views of the commander, and which can feldom fail to point on fome diftant, and diftinct object, that will ferve to correct the pofition of the different perfons who fucceffively as their feparate bodies require it, prolong the line from the feveral points already eftablifhed in it.——
In general therefore the point (A) where a formation or entry into an alignement is to be made being marked by a fixed perfon, the commander will place a fecond (o.) 30 or 40 paces, without the firft, exactly in the direction which he determines to give to his new line, and which will generally be on fome diftant object. Thefe two perfons will mark a bafe, which by adjutants (a. a. a.) or others fucceffively aligning themfelves backwards on the two firft placed men, and on each other, may be prolonged to any required length, at the fame time that the diftant point (D.) ferves for the commander who perhaps alone knows it to correct them upon

<small>Fig. 13.</small>

<small>Method of afcertaining points of movement or formation.</small>

3. Two original or bafe points (o. A.) which are to be prolonged or formed upon, fhould not be too clofe together, otherwife the direction

tion of the line must be indistinct, and the farther they are asunder the better can a line be taken upon them.—Where two points (o. A.) are to be given in a certain direction towards a distant one (D) the innermost (A) should be first determined, and the outer one (o) is immediately and easily taken over the innermost, and the distant one (D) of correction.—Should the outer one (o) be first taken, time is lost in directing the shifting of the inner one (A.) before it is truly lined on the more distant point (D); besides the point (A.) in many changes of position of a line or column is naturally the first ascertained (being the pivot flank of a company on which the change is to be made, or the point of march towards which the column is moving) and from thence the distant point (D) is then taken, which gives the new direction, and depends on the eye and intention of the commander; the easy ascertaining of (o) follows of course.—Or the commander after ascertaining (A.) will fix (o.) *ad libitum*, and find out (D) if such object presents itself in the prolongation of the other two.—At any rate (A.) is the point first to be determined on.

Fig. 13.

―――――

4. When the persons who prolong a line are on horseback, the head of the horse of each standing perpendicular to that line is the object, and when they dismount their own breast is the object, which the shoulders of the leaders of the divisions of a column in march, rase in passing, and which is in the line of the head of the horse.—It is also the breast of such other men, as may be posted on foot, which the several leaders in like manner rase, as they successively arrive at them.

Position of prolongers of lines.

Although the leaders of the two first divisions of a column march on the persons placed in the line, yet if its direction happens to be on

some

some remarkable object, they should as soon as possible discover it, or be shown it as the general correction of the march.

Fig 14.

Method of prolonging a line by officers or serjeants.

5. When a number of officers or serjeants (s. s. s. s.) are to be individually, successively, and separately advanced in order to give a direction on which pivots of the divisions of a column are to stand; or flanks of divisions which successively come into line are to be halted, or on which the dressing of a battalion is to be corrected.—Two such persons will be truly, and previously placed, and the others the more exactly to attain a perfect line, instead of attempting at once to dress by each others breasts, will first cover in FILE with precision at their required distances, and then carefully front as directed, before their several divisions move up to them.—Were such persons to endeavour to take up their ground at first, by dressing in a line; the least inclination backward or forward of the body, and the certainty of the shoulders turning, when the eyes are directed to a flank, would make it a difficult operation: But in FILE when each places himself square on the line, covering the necks of those before him, the inclination of the body backward or forward does not affect the direction, and the end proposed is at once attained.

Fig. 12.

6. In successive forming of divisions into line, as from close column, from echellon, &c. the first division (A. a.) that arrives in, and is truly formed on it may be considered as the BASE which is constantly prolonging for the others; the men as they come up endeavour to line well on the part already formed, and the officer corrects that lining on the distant point in the true prolongation which is prepared for him by his adjutant or other persons, just beyond where the flank of his battalion is to extend, and thus battalion after battalion arrive in line.

7. The

7. The ascertaining of the points necessary for the movements and formations of the battalion is the particular business of the adjutant in the field; and in this in exercise he may be assisted by two detached persons placed behind each flank of the battalion, who are properly trained, quickly to take up such line as he shall give them; but for this purpose they are not to run out before their aid is wanted, nor are they to make any unnecessary bustle, and when the operation for which they were sent out is accomplished, they will immediately return behind their proper flank.

S. 85. *Dressing.*

1. In DRESSING when halted, a small turn of the head is necessary, and is allowed in order to facilitate it.——When the word *Dress* is alone given, it means to the hand to which the troops are then looking, and when eyes are at the same time to be turned to a new point in order to dress, it should be so expressed by the addition of *right*, *center*, or *left*.——But whenever the word *Halt, Dress*, is given by an officer to his division, it always implies that the men are looking, or are to look to such officer, who is then on the flank of Appui.

2. All DRESSING is to be made with as much alacrity of officer and soldier as possible, and the dresser of each body as he accomplishes the operation will give a caution *Front*, that heads may then be replaced, and remain square to the front. If the body to be dressed is extensive,

as that of a battalion or parade, the dresser must justly place one division before he proceeds on that which is beyond it.

Fig. 15.

General attentions of dressing in all formations.

Fig. 15.

3. On all occasions without exception of FORMING and DRESSING in line, it must be remembered that the soldiers come into line with their eyes directed to the general point of *Appui* (A) where the leading flank is to rest, and of course towards whatever part of a line is nearer that point than themselves, which may be already formed before them, and is to them a direction.——But the officer in dressing (without exception) is placed on that flank of his division or body towards which the mens eyes are turned, and from thence he makes his corrections of the other flank on the distant point (D) which is previously marked by the adjutant, or some other person placed in the true general line: therefore on all occasions by the mens lining themselves to one hand, and the officers correcting to the other, the most perfect line may be obtained.——Should it be neglected to give or prepare such points of correction, the dressing of the line would be irregular, and slow, and depend entirely on the men taking it up from each other, and from the first formed flank, which is an imperfect method, and can never produce a just line, capable of marching forward in due order.——The having such points quickly and successively prepared the instant before they are wanted, and without any noise, or apparent bustle, so that no delay may be made in the operations of the battalion or line, is one of the great attentions of the commanding officer and adjutant, to which also the intelligence of the trained persons placed behind the flanks will much contribute.

4. If

4. If the open column is to enter on an alignement, there must be three prepared points; one (A) where it enters, and which serves as a future point of correction in march, and in forming; and two more (a. D) always advanced before it.

Points necessary to be given in movements, or formations.
Fig. 13.

5. If the close column is to form in line on a flank division, it must have a point (D) of march and correction beyond the other flank, and intermediate small points must also be taken by the leader of the front division, in order to preserve its direction of movement. If it forms on a central division, it must then have a point of correction to each flank (D. D.) and march justly on intermediate ones.

Fig. 16.

5. If a battalion takes a new position by the echellon march, there must be a point (A) given where its leading flank enters the line and forms on it, and another (D) just beyond its extreme flank on which the dressing of each division is corrected.

Fig. 12. 15.

7. If the battalion changes position to a flank, by the filing of divisions, the prolongation of two points (A. o.) given in front of the pivot flank of its leading division will determine the direction of the other pivots.—If the change is central, one central point (a.) and one (o. o.) on each side of it being prolonged will determine in like manner the line of the other pivots.

Fig. 14. 18.
Fig. 17.

---

8. The commander will himself generally have a distant point, on which he will determine those battalion points, and which will serve him as his point to correct the whole.—Independent therefore of the partial helps which advanced serjeants may give to the formation and

dressing of their several divisions, it may be observed with respect to one or more battalions, that in marching in front, or in column *two* advanced points and *one* rear point are necessary: and in successive formations into line, besides its point of *Appui* which each body moves up to, *one* distant point taken in the determined direction, and beyond where the battalion is to extend, is essential for the correction of its dressing, and in this line is every division exactly brought up, and dressed.

OPEN COLUMN.

Formation of columns.

1. All COLUMNS are supposed formed from line for the convenience of movement, and for the purpose of again extending into line.—Every column of march or manœuvre must be formed by a regular succession of the divisions from right to left of the line, or of such of its parts as compose the column, for whatever is the relative position of a body in line, such ought it to be in column: and where several connected columns are formed, the same flanks of each should be in front, but whether Rights or Lefts will depend on circumstances.———Columns formed from the center of battalions or lines, should seldom be made, are partial, and not adapted as the others are to movements and formations in all situations.

Columns of march and manœuvre.

2. The chief objects of the OPEN column are, facility of movement; the quick formation of the line to the flank, and the change of situation

in

in the shortest lines from one position to another.—It is named the column of march or ROUTE, when applied to common marches, where the attention of men and officers, are not so much kept on the stretch.—It is named the column of MANOEUVRE, when being within reach of the enemy the greatest exactness is required in order to its speedy formation at any instant into line during its transition from one position to another.

3. Columns of march or manoeuvre will generally be composed of companies or subdivisions.—For the purposes of movement they need not exceed 16, or 18 files, nor should they be under 6 file in front, when the formation is three deep, otherwise there will not be space to loosen the ranks, and the battalion will of course be lengthened out.—An open column occupies the same extent of ground as when in line, minus the front of its leading division: But a body obliged to march any distance in file, will at least occupy one half more ground than it requires in line; such situation is therefore to be avoided. Front and extent of column.

4. From line the column is formed, and marches to the front, flanks, rear, or in any intermediate oblique direction, with either its right or left in front.—In each case the battalion or line WHEELS the quarter circle by divisions to either flank and HALTS. The whole MARCH.—The leading division *wheels* into, or moves on, in the prescribed direction, and the rest follow in column. Formation of the open column from line.

5. The open column, or the column at half or quarter distance, may also be formed oblique or perpendicular to the line, on any given division; by the other divisions (according to which flank is ordered to lead) wheeling, filing, and placing themselves in front, and rear of the given one.

6. Columns

6. Columns of march or manœuvre will be formed with the left in front, whenever it is probable that the formation of the line will be required to the right flank; and *vice versa*, when required to the left flank.

BATTALION OPEN COLUMN.

Dressing in column.

Fig. 23.

1. In column divisions cover and dress to the proper pivot flank: To the left, when the right is in front: and to the right, when the left is in front.——The *proper* pivot flank in column is that which when wheeled up to, preserves the divisions of the line in their natural order, and to their proper front: The other may be called the *reverse* flank.

Distance of ranks.

2. In column rear ranks (if not ordered to be locked up) are one pace asunder.—When a considerable distance is to be marched, they may be opened half a pace more, but without encreasing the distances of divisions, which remain such as are prescribed according to the object of the movement, and which are always taken from front rank, to front rank.

3. The post of commanding officers in column, is each near the flank of the leading division of his battalion.

Leading officers.

4. Each division of which a column is composed is conducted by a leader placed on its pivot flank of the front rank which is his general post.—

post.—In a column of companies or platoons such leader is the platoon officer.—In a column of sub-divisions the officer leads the head sub-division of his company; and his covering serjeant in battalion the second.—In a column of sections the platoon officer leads the head section of his company; his serjeant the middle one; and an officer or serjeant from the rear the last one.——When divisions are filing from column into a new position, their several leaders conduct their heads.—When any considerable continuation of the march is the object, and that pivot officers are permitted to be in front of their divisions, their flank posts must be occupied by non-commission officers, who remain answerable under their direction for the preservation of the proper distances.

5. In open column, the artillery, music, drummers, &c. of battalions wheel with and remain closed up to the rear of their respective divisions.—In column at half or quarter distance, they may occasionally if there is space move in file, on the flank which is not the pivot one.—Instead of being kept collected, they may in column of march be sent to their respective companies to remain in the rear of each: But on no occasion whatever is the assembling of them to be allowed to lengthen out, or interfere with the movements of the battalion or column, or to encrease the intervals betwixt battalions in column.

Music, drummers, &c.

6. On all occasions of wheeling from line into open column (except where the narrowness of ground prevents it) the divisions WHEEL BACKWARDS on their pivot flanks.—The advantage so great, and the necessity so evident of having the *pivots* remain covering each other truly, as well as having just distances preserved is thereby secured, which will never be the case in *wheeling forward*, from the different

Fig. 21. 59.

strength

strength of battalions in a line, and of companies or divisions in the same battalion.——In wheeling backward if divisions do not exceed 16 or 18 file, they may readily WHEEL back without facing about, but if divisions are stronger and the ground uneven, they must FACE about—WHEEL—and then *Halt, front.*

<small>Wheeling forward into line.

Fig. 25.</small>

7. When an open column is to form in line to its proper front, the divisions will always WHEEL FORWARD on their pivots: But should it be meant to reverse the front, the PIVOTS themselves must then wheel forward, which will prevent any false distances, that unequal divisions would occasion, although the flanks they do wheel upon, may not then be in a regular line.—Should the divisions of the column be of equal strength the front may then be reversed by wheeling back upon the pivots which will preserve the regularity of the alignement.

<small>Wheeling on the center.</small>

8. Platoons must be accustomed to wheel occasionally upon their CENTER, half backward, half forward, and to be pliable into every shape which circumstances may require, but always in order, and by the decided commands of their officers.

<small>Filings.</small>

9. All marches of battalions are made in column of companies, or other divisions, never by files where it can possibly be avoided.——Filings are only applied to the internal movements and formations of the divisions of the battalion and in some changes of position, not to any considerable manoeuvres of the entire battalion, or of greater bodies.

<small>Wheelings in column.</small>

10. All wheelings, and filings made from the halt, from line to form in column, or from column to form in line; are made at a quick step.

11. When

11. When the rear ranks clofe or open on the march, in the one cafe they will ftep nimbly up, in the other they will flacken their pace until the due diftance is attained.—In both cafes the front rank continues to proceed at its then rate of march.

Opening or clofing of rear ranks.

12. In an open column of manœuvre of one or more battalions the divifions ought as much as poffible to be equalized.—The whole muft be put in march at the fame inftant and the ftep preferved, equal as to time and length whether marching on level or inclined ground.—Every divifion muft trace out the exact track which the leading one does; nor muft any part make a partial alteration of pace.—Thefe circumftances obferved which will preferve the juftnefs of wheeling diftances and the covering of pivot flanks; and no embarraffments being allowed in the intervals of battalions, an exact line to the flank is at any inftant procured, by the wheel of the quarter circle; and all clofing in, unfteady fhifting, and after dreffing is avoided.

Peculiar attentions in the open column of manœuvre.

13. The countermarch by files of the divifions of a column each on its own ground; changes a column that is ftanding with its right in front, into a column with its left in front, and thereby enables it to return along the ground it has gone over, and to take new pofitions without altering or inverting the proper front of the line. (S. 53. 100.)

Fig. 40 B.
Countermarch by Files.

14. The countermarch by divifions fucceffively from the rear to the front, changes the leading flank of the battalion column, but allows it to continue its former direction of march, and is a previous manœuvre often neceffary and required to enable a battalion to take up a relative pofition. (S. 101.)

Fig. 41. 42.
Countermarch by divifions.

f

All

All counter marches neceffarily change the pivot flanks of columns.

<small>Fig. 22. 24.
Wheels in Column.</small>

15. Open column of companies will in general wheel on a *fixed* pivot, except that in the continuation of a march, they have fufficient ground gradually to make their changes of direction on a moveable one, if fo ordered.——Columns of fubdivifions or fections will always wheel on a *moveable* pivot when it can be done.——Columns at half or quarter diftance muft alfo make their neceffary wheels on a *moveable* pivot, otherwife a ftop muft enfue.

<small>The front of column not to be altered when marching in an alignement.</small>

16. No doubling up, encreafing, or diminifhing the front of the column muft be made, after entering on a ftraight alignement, in order to form in line. Such operation when neceffary fhould be performed, before the line of formation is entered on.

<small>Fig. 24. 25.</small>

17. In whatever manner the leading divifion of a battalion column arrives in a ftraight alignement on which it is to form, a mounted officer always gives the point where it enters.—And when arrived at its ground, that it halts and is to form, the commanding officer from that divifion corrects if neceffary the pivot files on the fixed diftant points, before the divifions wheel up into line.

<small>Pivot officers.

Fig. 24.</small>

18. Pivot officers of columns when marching in an alignement, muft be fteady on the flanks of their divifions, as they give the true wheeling diftance, and covering of the pivots in their own perfons: They muft not look to or endeavour to correct the march of their divifions, that care muft be left to ferjeants, and other officers in the rear. The pivot files of men (that they alfo may be truly covered when halted), muft be clofe to but not touch, or derange their leaders in the march.—
The

The pivot files of the open or close column in march are always directed and conducted on the given points of march whether the column is moving in a line on which it is to form, or whether it is moving up to a point where it is to change its direction; and the leading officer in column always leaves the object on which he has marched, or at which he wheels, close on his own outward hand.

19. When marching in a straight alignement there must never be more than one officer (or leader) on the pivot flank of each division, all others are either on the opposite flank, or in rear of the divisions.— Nor are such leaders then covered in the rear ranks by their serjeants, in order that they may the more easily see and distinctly cover, each other in the given line. *Officers and colours in the alignement.*

The colours cover the 3d files of men from the pivot, and must be ready to move up, when the line is to be formed.

20. In marching in an alignement on advanced points, such points must be known and visible to the leader of the second as well as of the first division; because such second leader must preserve his first, and the given points in the true line, and on the accuracy of the position of those two leaders, depends the covering of the rear ones.—Officers who have an indistinct sight, can never lead the two first divisions of a column marching in a straight alignement, and must therefore on such occasion be replaced by other persons, whose accurate vision enables them to preserve and prolong the just line which the whole are to follow. *Officers that conduct the leading divisions of a column must not be short sighted.*

21. All marching in the alignement must be made in ordinary time, and taken up before, or from the point where it is entered with precision, the pivot officers are then peculiarly answerable for distances, *Marching in an alignement.*

f 2 and

and exact covering of the flanks.—To march with accuracy in an alignement in quick time, so as at any instant to be ready to wheel up into line, and (without a considerable pause) to move on, is an operation hardly to be expected, and seldom to be required.

<small>Wheeling into line from open column.</small>

22. When the column of companies halts to form: pivot flanks are in an instant corrected from the leading division by commanding officers of battalions.—Leading officers move into the front of their platoons.—Their covering serjeants place themselves on the right of each if the wheel is to be to the left; or otherwise behind the pivot file if the wheel is to be to the right.—Pivot men of the front rank face square into the new direction.—The whole wheel up, and halt.—Officers dress the interior of their platoons, and then replace their serjeants who are now in the front rank.—If any farther dressing is necessary, it must be ordered and made by a mounted field officer.

<small>Dressing on pivots.</small>

23. If the battalion after wheeling up from column into line, is not critically well dressed, the fault must be in the internal parts of the divisions: This must be immediately corrected (by each platoon officer) on the pivot men, who on no account must move, or shift, but remain so many given or fixed points on whom the battalion is exactly lined.—Each platoon officer thus only dresses within his own platoon; if a more accurate dressing is required, it is afterwards given by a field officer.

<small>Forming in line.

Fig. 25.</small>

24. In general the whole of a battalion, will be halted on its ground, stand in column, and its pivots be adjusted, before it wheels up and forms: but if necessary and where parts of it arrive in the line by *filing*, they may form successively as they come up.—If part of a battalion should therefore be ordered to wheel into line while the other divisions are

are not yet in it, the pivot men of thofe divifions (and not the officers) muft cover on the formed part of the line before they wheel up.———— And when feveral battalion columns changing pofition enter feparately and are to form in the fame line: each may be fucceffively wheeled up, if fo ordered or intended, when its adjoining one has three or four of its divifions ftanding in column on the line.

25. When a point of entry is marked in a new alignement the *pivot* flank of the leading divifion of a column is always directed on fuch point.—If the line is to be formed, and the head flank placed at fuch point of entry; the head divifion will reft its pivot on the line and at a diftance equal to its front from fuch point.—If the rear flank is to be at that point; the pivot of the rear divifion will halt at it.—If the point is an intermediate one; a central divifion will halt at it.—The line will be formed by the wheeling up of divifions, when they are feverally placed upon it.

26. On fome occafions (as in paffing lines, forming clofe columns, &c.) the platoons or divifions of a battalion in line are ordered to FILE to front, rear, or into column without firft wheeling the quarter circle.— An explanatory caution being given; at the word for the battalion to FACE, the platoons face to the point directed, and at the fame inftant the three leading files of each throw themfelves to the flank according as they are to move, fo as to be difengaged from the laft file of the preceding platoon.———In this fituation each leader is enabled at the word MARCH, to move independant, without check, and on his proper point. *Difengaging heads of files.* Fig. 36.

27. The rear divifions of a battalion or more confiderable column in march, conftantly follow every turning and twift which the head makes; *March in column.*

makes; each successively changing its direction at whatever point the leading division may have so done.—When at any accidental moment the column is ordered to HALT, and FORM in line; the pivot men of platoons must remain steady where they are found at the word HALT, and the divisions will wheel up into what will probably be a curved, but a just line. If the march in column is again to be resumed, the line breaks backward, and the rear divisions at their ordered distances will continue to follow the exact path traced out by the head; nor are the following divisions of a column ever to deviate from this rule, or endeavour of themselves to get into a straight line when the general direction is a winding one, until an express order is given for that purpose; which can hardly ever be the case until the head of the column is halted with a determination to form the line in a straight direction.

March of the column thro' a wood, or in embarrassed ground.

Fig 26.

28. The march in column through a thin wood, or in ground where impediments frequently change the direction of its head, or along the winding of heights which are to be occupied, will be best made by subdivisions, or by sections of five or six file in front.——The *pivot* files will preserve exact distances from each other, choose their own ground, and wind as the trees or other impediments permit, along a general direction: When the column *halts* and forms, the line will be a continued curve, which can afterwards be easily made straight, if circumstances require it.—In such situations, at no time if it is possible should any of the *pivot* flank leaders be obliged to double or quit the continued line of march; but the other files may be (when impediments are to be passed) much opened or loosened from those pivots, who in the mean time moving free and preserving wheeling distances, are in a situation at every instant to *halt* and form in line, the others closing into them.

29. Should

29. Should the march in a ſtraight alignement be at any time inter- *Obſtacles in march in an alignement.* rupted by pools of water or any other obſtacle which is impaſſable, the march will be continued ſtraight to that obſtacle, the obſtacle will be ſurrounded, (and always if poſſible by deviating to the reverſe flank ſo *Fig. 27.* as to remain behind the line) and the ſame ſtraight line will again on the other ſide be taken up by the pivots, at the point in it which a detached perſon has prepared.—Allowance will be made when the line is to form, for the breadth of ſuch obſtacle, by the doubling of as many diviſions as will fill up the vacancy (when it can be done) which is thereby occaſioned in the line; nor muſt any ſmall interruptions in the line that can poſſibly be ſurmounted, ever make the pivots deviate from the ſtraight line, when the intention of forming on the line is evident and known to all.

ASSEMBLY OF THE BATTALION, AND GENERAL CIRCUMSTANCES OF EXERCISE.

The companies having been inſpected by their officers on their parti- *Aſſembly of the battalion.* cular parades ſhould arrive and ſtand on the parade of the battalion in open column of companies and with either right or left in front.—The ground is given by which ever diviſion firſt arrives on it, and the others arrange themſelves in front or rear accordingly.——In this ſituation; are reports made to the commanding officer; companies equalized; muſic, drummers, pioneers, &c. aſſembled at their proper ſtations; all other individuals of the battalion placed; pivot files, and juſt wheeling diſtances corrected.—The battalion is then formed in

line

line by wheels of the quarter circle, and by word from the commanding officer; the colours are sent for and posted; and the whole are thus in readiness to move, by subdivision or company column.

<small>March to the ground of exercise.</small>

The march to, and from the field in column, should be considered as one of the most material parts of exercise, and be made with attention, equality of step, just distances, and perfect order.—The front of the march should be frequently encreased and diminished in the manner prescribed (S. 87.) and the battalion at different periods formed by wheels to the flank, to show that distances have been duly preserved.

<small>Exercise by companies.</small>

<small>Fig. 19.</small>

The exercise of the battalion must frequently be preceded by that of companies in detail according to the instructions given in the second part.—Therefore when the battalion is arrived at its ground, the officers will be assembled, and those commanding companies informed what particular parts, (referring for this purpose to the numbers marked in the exercise of the company) in what succession, and for what length of time, or how often each operation of the company is to be repeated.—The companies will then by a regular process be separated, by taking intervals in one line, or in two lines, so that each shall have a free space of 40 or 50 yards square.—They will on that ground begin and finish in nearly the same instants of time, each of the ordered points of exercise.

S. 86. *Exercise of the Battalion.*

The above being accomplished, the companies will be ordered to assemble in line, or in column, and the BATTALION again united and

and formed will proceed to its particular EXERCISE as contained in the following articles, which may be claffed and arranged according to circumftances, and the views of commanding officers: the modes of execution being detailed hereafter under their proper heads.

Detail of exercife by the battalion.

The Battalion ftanding in open Column.

Fig. 74. { 1. The column will clofe to half, quarter, or clofe column, and again take open diftances either from the front, or rear divifion. S. 153.

Fig. 40. B. 2. The companies may fingly countermarch by files. S. 100.

Fig. 41. { 3. The flanks of the column may be changed, by the rear company becoming the front one, in confequence of a countermarch of the whole column from the rear. S. 101.

Fig. 49. { 4. The pofition of the column may be changed to either flank, by the companies facing, filing into the new direction, and halting with their pivot flanks on it. S. 123.

Fig 51. 54. { 5. The open column will form in line.—By wheeling up when the whole is in the alignement.—By halting the head divifions in the line; filing the rear ones into it, and then wheeling up the whole.—By the head divifion halting on the line, and the rear ones wheeling back into echellon pofition, after which they move up into line. S. 118. 124. 127.

Fig. 47. 48. { 6. A front, center, or rear divifion of the column may be placed in a new given direction, and the reft by file marching will take up their ground. S. 120. 121. 122.

7. The

               7. The column at half or quarter diſtance will form.——By filing into line.—Obliquing into line.—Diviſions wheeling ſucceſſively into line and taking open diſtances.

               8. The diviſions may face to either flank, march the lock ſtep, halt, and again front into column. S. 123.

               9. The cloſing ſtep may be practiſed by the whole column at once. S. 43.

Fig. 28, 29. 30.   10. The front of the column in march will be encreaſed, and diminiſhed, and the column will occaſionally wheel to the flank into line to ſhow the preſervation of diſtances. S. 87.

Fig. 59. 13.   11. From line the companies, or other diviſions may wheel backwards on their pivots into open column, and to either hand. S. 108.

Fig. 52.   12. March and prolong the line to the flank. S. 115.

Fig. 49.   *Wheel up into line.* S. 118.

Fig. 47. 48. 80.   13. Change of poſition on a central or flank company by filing, or by the echellon march of companies. S. 120. 159.

              14. The battalion may march in file to the flank at the lock ſtep, and front. S. 94.

              15. The battalion may take 20 or 30 ſide ſteps to the flank without opening out. S. 43. 79.

              16. The battalion may advance in line, and halt. S. 166.

              17. The battalion may retire in line, and halt front. S. 168.

                                                           18. The

[51]

18. The alternate companies will form two lines and march to front, and rear preserving intervals. S. 175.

Fig. 46 — 19. Passage in file through a second line, or wood, to front or rear, from a flank of each company. S. 174.

20. Passage of the obstacle in the march of the battalion, by divisions doubling as ordered. S. 170.

21. The oblique march of the battalion, and change of direction by gradual alteration of the shoulders. S. 169.

Fig. 14. — 22. The battalion halted to be dressed, by advancing the platoon officers, and moving up the men. S. 167.

23. The whole or a wing of the battalion to be thrown forward on the center or flank, by placing a few files, and the rest turning their shoulders, and gradually dressing up.—The same done backwards gradually at a short step without facing about.—Eyes being directed to the point of forming on all occasions.

Fig 75.76. — 24. The battalion will advance, and retire in echellons of companies.—Form in line on any named one.—Throw backward or forward any number of companies into echellon.—Wheel them into oblique line.

25. The battalion retiring in two lines by alternate companies, may make a degree of wheel during the movement, so as to give a new direction to the line.

Fig 31.33. 35. — 26. The battalion may pass a defile or bridge, to front, or to rear. S. 91. 92.

g 2   27. The

[52]

Fig. 39. { 27. The battalion may countermarch by files from one to the other flank.—Alſo upon the center from both flanks.——Alſo from, and upon the center. S. 97. 98.

Fig. 40. A. { 28. The battalion may countermarch by diviſions from one to the other flank.—Alſo upon the center from both flanks. S. 99.

Fig 37. 38. { 29. March of the battalion by diviſions from one flank towards the other, either behind or before the front, each diviſion wheeling and following ſucceſſively the one that precedes it. S. 96.

{ 30. The battalion from line forming the ſquare or oblong, marching, and again forming in line.—Or from the ſquare marching off in double column through a defile.

In cloſe Column.

Fig. 64. 65. 66. { 31. The cloſe column is formed on any named company. S. 137.

32. The direction of the cloſe column is changed. S. 141.

Fig. 74. { 33. The cloſe column is opened out from the front or rear, and again cloſed up on any diviſion. S. 153.

Fig. 70. 71. 72. { 34. The cloſe column of two companies in front is formed from the column of one company in front. S. 147.

Fig 67. 68. 69. { 35. The line formed either from the column (of one or two companies in front) halted, on a front, rear, or central diviſion.—Or from the column moving in file to its flank, on a front or rear diviſion.—Or by an oblique deployment of its diviſions. S. 144. 148.

36. The

{ 36. The exercife of the firelock in all its parts both by companies and battalion, and efpecially loading and firing.

The FIRINGS may be applied and intermixed with thefe movements as found proper, and fuch other circumftances of formation and exercife as fpace allows of, and as occur to commanding officers, may agreeable to the eftablifhed modes laid down, be from time to time executed.—But the above have been more particularly felected, as including almoft all the various movements that can be required in the operations of the battalion when fingle, or united with others in line: They may be combined according to the ground, and to the views of the commander, and may arife from different fituations by altering or adding the connecting circumftances, and the particular detail of their execution is to be found in the fections referred to.

The light company and grenadiers are generally fuppofed acting in line with the battalion: But the light company may be occafionally placed half of it behind each flank of the battalion; in that fituation it is ready to cover the front, rear or flanks of the column when in march; to protect the forming of the line; or to cover its retreat.—For thefe purpofes, it may from time to time be detached and act in divifion or individually as circumftances may require, and in the manner fpecified in its particular exercife.—It can feldom be obliged to run or hurry; in fuch cafes as demand it, it will march quick but in order, with files loofe but not too open, and always under the command and guidance of its officers.

Light company.

Mode of instruction.

On all occasions of common parade; a guard, a battalion, or its parts should never assemble, or be dismissed without performing some one operation or other of movement, and of the firelock.——In this manner by simple, and imperceptible practise, the steadiness, and instruction of every individual is attained, and officers become perfect in the three great and important field duties of precision and energy in their commands; exact distances of march; and the correct dressing, and covering of pivots.—The time often unnecessarily consumed in the field in detail and manual exercise will also be saved, and the battalion be there solely employed in executing the prescribed movements applied to such circumstances, and varieties of ground as present themselves to the commanding officer; the modes of execution being already thoroughly understood, and instantly applied by each individual.

———

Attentions in exercise.

Fig. 20.

Single companies or battalions when at exercise must generally consider themselves as part of a line, and not always as detached, or independant bodies: Their movements and formations should be on a supposition of lining with other troops already placed on their flanks.—— Two or more persons separated at a proper distance from one another, and from the company or battalion, may represent the flanks and center of an adjoining battalion, and may always first take their station in the new line. This would cause the formations to be made on determined, not on accidental points, the practise of which latter usage much tends to occasion that incorrectness and deficiency which sometimes appears when any number of our battalions are directed to move, act, or form, in concert.——In general the battalion should not be looked on as a perfect or separate body, but only as a member of the line;

line; its movements as relative to and dependant upon those of others, and its principal operations should be calculated accordingly.

In exercise the two flank companies may be occasionally separated Fig. 20. from the battalion, and represent the center of two other battalions; one of them will be named as the directing one in march, and the halt and dressing of the battalion, will be made from its own center towards each of them as is directed for a line of battalions.

Diminishing or Encreasing the Front of the Column.

The column of march or manœuvre in consequence of obstructions in its route which it cannot surround, is frequently obliged to diminish its front, and again to encrease it, when such difficulties are passed; it is one of the most important of movements, and a battalion which does not perform this operation with the greatest exactness and attention so as not to lengthen out in the smallest degree, is not fit to move in the column of a considerable corps.

The encrease or diminution of the front of the column is performed by the battalion, when in movement or when halted.——In movement this operation is either done by each company successively, when it arrives at the point where the leading one of the column performed it;

or

or elſe by the whole companies of the battalion at the ſame moment.—In either caſe the chief of the battalion at the inſtant that it ſhould begin to reduce or encreaſe its front, gives the general CAUTION ſo to do, and the chiefs of companies give their words of execution to the ſubdiviſions or ſections to double behind, or move up quick to the regulating ones which preſerve their original diſtances from each other, and never alter the pace at which the column was marching, but proceed as if they were totally unconnected with the operation that the others are performing.

When the column of companies is to be reduced to that of ſubdiviſions or ſections, it will always be done by the others doubling from their pivot flank, behind their reverſe flank ſubdiviſion or ſection, ſo that the battalion may remain ready to form in line by a ſimple wheel up to the flank; therefore the doubling will be behind the right when the right is in front, or behind the left when the left is in front. ——When the front of the column is to be encreaſed; the ſubdiviſions or ſections that doubled will move up to their leading one by a quick incline.——As in diminiſhing or encreaſing the front of the column in march, the pivot diviſion is the one that quits its direction, the exactneſs of pivots after ſuch operation will appear to be interrupted; but this is of no conſequence and inſtantly regained in a column of march; it can hardly ever take place in a column of manœuvre which has entered a line on which it is to form.

S. 87. { *When a Battalion Column of Companies in march diminiſhes its Front, either by Companies ſucceſſively, or the whole Battalion at once.*

When the leading company arrives within 12 or 15 paces of the point where it is neceſſary to diminiſh its front; the commander will give a lond CAUTION that the ſubdiviſions are to double either by companies ſucceſſively, or the whole battalion at once.

Fig. 6. A.   *If ſucceſſively.*——The leader of the head company proceeds as directed (S. 57.) and each other does the ſame when it arrives on the ſpot where its preceding one doubled.

Fig. 6.   *If at once.*——On the general CAUTION from the battalion commander, each company leader, without waiting for each other, proceeds as directed. (S. 57.)

S. 88. { *When the Battalion Column of Subdiviſions in march forms Column of Companies.*

The battalion commander gives a loud CAUTION, that column of companies are to be formed either ſucceſſively, or by the whole battalion at once.

Fig. 6. B.   *If ſucceſſively.*——Each inclines up as directed (S. 57.) when its leading ſubdiviſion arrives on the ground, where its preceding one formed up.

Fig. 30.   *If at once.*——On the general CAUTION from the battalion commander: Each company leader proceeds as directed (S. 57.) without waiting for each other.

When divisions double back or form up in column, ranks must be closed, arms carried, and the transition from one situation to the other made as quick as possible; and as soon as the column is in its new order, the pivot flank leaders place themselves on those pivots.

When the front of a column is to be diminished, and the obstacle is before the part which is not to double, such part must incline after the doubling is made in order to pass it: but timely attention is to be given, to bring up if possible by inclining, the part which is not to double, square to the opening through which it is to pass, before such doubling begins.——And when a diminution of front is immediately to follow an alteration in the direction of the march, such alteration should be made with a gradual sweep, so as to give the head of the column its new perpendicular direction, when at least 12 or 15 paces from the point of breaking off.

The successive breakings of each division of the column at the point of difficulty, and its subsequent moving up again as soon as it has passed it, is the most general practise, but is the most likely to lengthen it out, which is the great evil to be avoided.——The reduction of front by the whole battalion at once, is therefore the most eligible;

and

and for the same reason, the encrease of front (when the rear of it has cleared the difficulty) by the whole battalion is to be preferred.

As in a confiderable column the fucceffive doubling, or forming up of companies would be performed by each when it arrived on the identical fpot where the leading one of the column doubled or formed up. ——So when this operation is done fucceffively by battalion, each will at once in the fame manner perform it when its head is arrived at the fpot, (and of which it muft be apprized) where the head of the preceding battalion was, when it fo doubled or moved up.

S. 89.
Fig. 5. B.—28.
{ *When the Battalion Column of Companies is halted, and to diminish its Front.* }

CAUTION. { The chief will give the CAUTION to form column of fubdivifions or fections: on which the covering ferjeants will fall back and mark the future pivot flank of the doubling fubdivifions.

Inwards Face.
Q. March.
Halt, Front.
Drefs.
{ The leaders of each company will inftantly give the words *Face inwards* (difengaging their heads) *Q. March*; *Halt, front* to their fubdivifions or fections, when behind the ftanding fubdivifion or fection, and dividing juftly the diftances that exifted between companies; the flank leaders will then place themfelves on the pivots.

h 2　　　　　　　　　　　　S. 90.

S. 90.
Fig. 30. 5 B. { *When the Battalion Column of Subdivisions or Sections is halted, and to encrease its Front to Companies.* }

Caution.   The chief will give the Caution to form Companies.

To the—Oblique.
2 March.
Forward.
Halt, Dress.

{ The leader of each company immediately orders the bodies that move up, *To the—Oblique—2 March—Forward—Halt, dress*, when joined to the standing subdivision or section.—— The leader then places himself on the proper pivot flank of his company.

Should a column be retiring with the rear rank leading, the divisions will double as already prescribed so as to preserve the subdivisions or sections in their natural order for forming: and when the ground allows will again encrease the front of the column.

Fig. 29.   When the column has to pass a bridge, or short defilé, and that there is a certainty of immediately after resuming the front which it has diminished, then such part of the reverse flank of the leading division as the defilé will receive will pass it in front, and such part of the pivot flank as is necessarily stopt will by command *Face inwards* and follow close in file; on quitting the defilé the filing part will form up at a lengthened step, but the general rate of the column will at no time be altered: In this manner division succeeds division without any improper extension taking place. But if the column must continue any time

time on a reduced front, then it fhould fo be diminifhed by the doubling back of divifions.

When a clofe column, or one at quarter diftance is to pafs a defilé; before it enters, it muft ftand on fuch a front as will require no farther reduction; and therefore on approaching the defilé, a halt if neceffary muft be made, and fuch operation performed as will enable it to enter on fuch front as it can maintain in paffing.—When the defilé is paffed a new arrangement will determine the advance of the column.

Fig. 9 [?]

PASSAGE OF A BRIDGE, OR SHORT DEFILE, FROM LINE.

S. 91. { *A Battalion formed in Line may have to pafs a fhort Defile, or Bridge in its Front.*

Fig 91.

If before a Flank.—It will from that flank wheel into column, crofs on fuch front as will fill the defilé, and the column will be clofe or open, according as after paffing, it may be required either to deploy into line, or to prolong any given direction.

Fig. 93.

If before the Center.—The two center fubdivifions may ftand faft; the reft of the battalion will break inwards by fubdivifions; the whole will march forward in double column.—When paffed, the center fubdivifions ftand faft;

the

the others wheel to right and left, march to the flanks, and succeffively wheel up into line, (or) they proceed in march, and remain in double column 'till the head arrives at fuch point and is placed in fuch direction as the line is to be formed in.———Should the bridge or defilé only allow 6 in front to pafs. When the head of the double column arrives clofe at it, its two divifions having two paces diftance betwixt them, will file from their inward flanks to the front, pafs, and then move up into column as before, being in the fame manner followed by every other divifion.

But as many inconveniencies attend all central columns when a pofitive pofition is not to be occupied immediately after paffing a bridge or defilé; therefore in moft cafes the march in battalion column from one flank and on fuch front as is neceffary, is preferable; for from that order every poffible after fituation is accurately and eafily taken up; fuch as the windings of a height; the fkirting of a wood; or the prolonging of any given ftraight direction.

The battalion may alfo form clofe column of any given front, on the divifion which is oppofite the bridge or defilé, pafs in that fhape, and extend as ordered after paffing.

―――――

S. 92. $\begin{cases} \textit{A Battalion formed in Line may have to pafs a Defile} \\ \textit{or Bridge in its Rear.} \end{cases}$

If

If in the Rear of a Flank.——It will march off from the other flank behind the rear in column of companies or subdivisions successively, the front rank leading; wheel behind the standing flank; pass; and again wheel, and prolong any given direction.

Fig. 32.

If in the Rear of the Center.—It will march off as before from each flank, by columns of sub-divisions behind the rear, the leading ones when near meeting, will wheel inwards; pass in double column; and then if ordered, the divisions will wheel outwards successively and take up a line parallel to the one it quitted.—If the bridge or defile will not allow above six men to pass in front, the double column when it arrives at the entrance, will file to the front from its inward flanks as before directed; pass; move up into column; and either extend into line to each flank; or move on in any given direction.—In this way will the battalion be less liable to lengthen, than if it at once files from both flanks behind the center; passes; and again takes up its ground in file.

Fig. 35.

If after passing in files or columns, the march is to be continued forward; Should the wings be in file, they will form up to columns of subdivisions; the proper leading wing, according as the front of the line should be, will march on, and the other will follow it by countermarching its divisions successively from its rear, and in this manner the whole will be in column of subdivisions, which may be ordered if proper, to form companies.

The

Fig. 31.   The battalion may also form close column of a front equal to the breadth of the defile, behind or on the division nearest to it, and facing either way: it will then pass; and proceed according to circumstances.

S. 93. { *Where a Column of Divisions are successively to march off by wheeling from a Flank of a Battalion formed in Line, and that its Direction is towards the other Flank.* }

Fig. 34. A.   *If the Movement is made close along the Front.*—The leading flank division wheels up and marches along the line, and each other division successively wheels up behind it at the proper time, so as to follow in column, and to have its proper distance. Should the new direction make a small angle with the line, each division must move forward quick and successively to that direction, as its turn comes, before it commences its wheel, and so as not to lose its distance.

Fig. 34. B.   *If the Movement is made close along the Rear.*—The leading flank division wheels 3-4ths of the circle, and each other one successively half the circle, so as to have the remaining quarter to wheel, when its preceding division arrives at its pivot.—Should the line of march make a small angle with the old position, then each division after wheeling its half circle, will have to advance to that line in due time and successively,

successively, before it makes its remaining part of the wheel which brings it into column.

The open column may also in the before cases be advantageously formed by the successive FILING of divisions in the following manner, When the march is made from one flank of the battalion towards the other, and either along the front, or rear.

The leading division will *Face*; *March* out perpendicular to the line, its own length; *Halt Front*; *March*; and then proceed.—The division next it will *Face*, disengage its head towards the column; and when the leading division arrives, the other will then *March* quick in file; *Halt Front*; *March*, and thus follow division after division, each being ready and timing its several operations, so that the true distances are preserved. Fig. 36. A.

It is to be observed that in marches made in this manner along the FRONT the divisions face outwards or towards the moving flank of the battalion, and disengage their heads to the front: But in those made along the REAR, they face inwards or towards the standing flank and disengage their heads to the rear.—This method is peculiarly useful when the column moves to the rear, as much wheeling is avoided, and each division can with quickness and accuracy take its place in column. Fig. 36. B.

S. 94.

S. 94. March of the Battalion in File.

The march of the battalion in file and without opening out, can hardly be required except in smooth ground, and for the purposes of countermarching, or of closing, or opening an interval in line.

Face.
March.
Halt.
Front.
{ At the word Face, &c. the whole face to the hand ordered, and the officers take one side step to the front out of the rank, and are replaced by their serjeants.—At the word March, the whole step off correctly.—At the word Halt, the whole halt.—And at the word Front, they front, and officers and sergeants resume their places.—The officers being out of the ranks during the march (and which will take place whenever more than one company is to march in file) are of use in preserving the line and step.

S. 95. *General Formations of the Battalion from File.*

A battalion which has been obliged to move in file will form.

1. *To either Flank,* by halting and facing to right or left as is necessary.

2. To

2. *To the front* of the march by halting, facing to the flank; wheeling up by companies into open column, and then applying the formations of the open or close column to its required situation.—Or without halting and facing to the flank; the column of companies at once may be formed, by the files making a half face, and each marching up quick and diagonally to their respective leading men, who do not alter their pace; and as the pivot files are in the rear of companies, when they do come up, the column must be ordered to dress to them.

3. *To the rear* of the march, by first forming column of companies, and then applying the formations of the open or close column.

———

There can be few situations where the battalion must be formed to the front, or rear of the march, by the leading file halting and the whole moving up successively to it, and forming away in the rear of and beyond each other to one of the flanks.

The head of the battalion file must be so conducted as to leave sufficient space to the proper hand, for the other files to move up into open column of subdivisions or companies when ordered; and the pivot files in column are always the following ones, when the battalion is in file.—If the battalion is lengthened out when it is ordered to form;

it is evident that its facing into line, or its forming into column muſt be ſucceſſive as each file arrives at its place in line, or as each head file of the ordered diviſions arrives at a wheeling diſtance from the head file of the preceding one.

S. 96. *A Battalion ſtanding in narrow Ground may ſometimes be obliged to march in File, in order to form open Column from its leading Flank; either before or behind that Flank; before or behind its other Flank; or, before or behind any central Part of the Line.*

Fig. 37. A.
1. *If before the Right Flank.*—The right platoon will move on, the reſt of the battalion will FACE to the right, and MARCH in file; the diviſions will ſucceſſively *front*, and follow the leading one, and each other.

Fig. 37. B.
2. *If behind the Right Flank.*—The whole FACE to the right, and MARCH; the right diviſion inſtantly countermarches to the rear, *fronts* and moves forward, followed in ſame manner by every other diviſion, till the whole is in column.

Fig. 38. A.
3. *If before any central Point, or the Left Flank.*—The battalion makes a ſucceſſive COUNTERMARCH from the right flank towards the left, and when the right diviſion is arrived at the

the point from whence it is to advance in column, it again *countermarches* to its right a space equal to its front, then *faces*, moves on, and is thus successively followed by part of the battalion.—The other part of the battalion beyond the point of advancing, FACES inwards, when necessary makes a progressive march in file, and then *fronts*, and follows by divisions, as it comes to the turn of each; 'till the whole are in column.

4. *If behind the center or left Flank.*—The right part of the battalion COUNTERMARCHES from the right by files successively by the rear, and the other part of the battalion, as is necessary, makes a progressive march, by files, from its right to the central point, and there begins to countermarch: at that point the leading, and each other division *fronts* into column and moves on. Fig. 38. B.

When the left of the battalion is to be in front, the same operations inversely take place.

This METHOD of forming open column should only be used in narrow grounds, and in particular situations that require it, as in the passing of a bridge or defilé, or where the battalion stands in so confined a space as not to allow room for the wheeling of divisions.—The difficulties at all times of moving a large body in file, and the constant and unavoidable checks given to the equality and justness of the march by the divisions successively quitting the line, make it impossible in the above cases with due accuracy to take up the proper distances; and therefore whenever

ever the open column is to be formed from battalion and line, it ought to be done if possible by the wheelings of companies, subdivisions, or sections.

COUNTERMARCH BY FILES.

Fig 39 A.B   The *Countermarch* by *Files* is of two kinds.—Either SUCCESSIVE (the body being halted) by each file wheeling successively on its own ground as it comes to its turn: Or, PROGRESSIVE (the body being in motion) by each file wheeling, when it comes up to the point at which the leading file wheeled.—In the first case the body must shift its ground to a flank a space at least equal to its front: In the second it will perform this operation of the countermarch on its original ground, exchanging flanks and fronts; in both cases the pivots are in a small degree moveable.

The *Countermarch* by *Files* may be made either before or behind the body.—If made BEFORE it; the front rank men will be the pivots on which each file will wheel: If made BEHIND it; the rear rank men will then be the pivots on which each file will wheel.——All countermarches by file necessarily tend to an extension of that file; the greatest care must therefore be taken, that the wheel of each file is made close, quick, and at an encreased length of step of the wheeling men, so as not to retard or lengthen out the march of the whole, and unity of step is absolutely indispensible.

The

The *File* marching or countermarching of a battalion or greater body, will be made in ordinary time.—Of smaller divisions in general in quick time.

S. 97. *Countermarch of the Battalion, from both Flanks on its Center, by Files.*   Fig. 39. C.

A CAUTION is given that the battalion will countermarch. —The wings FACE from the colours which stand fast, and a serjeant remains at the point of each wing in order to mark the ground.—At the word MARCH the right wing files successively, close behind the rear rank and the left wing before the front rank of the battalion till they arrive at the points where each other stood.

    The BATTALION WILL COUNTERMARCH.

    BY WINGS OUTWARDS FACE.

    MARCH.

They then HALT, and the front rank of wings is quickly covered on the colours which have kept their ground, and served as a pivot on which the battalion turns.—The wings when covered in the line—FRONT, looking to the colours, and the colours take their places.—If a more accurate dressing is necessary, it must be given by the commanding officer.

    HALT.

    COVER.

    FRONT.

S. 98. *Countermarch of the Battalion from its Center, and on its Center, by Files.*   Fig. 39. D.

The

Caution. By Wings in- wards Face. By Wings, 3 Side Steps to the Right. March.	A Caution is given that the battalion will countermarch.——The whole face to the colours, which stand fast, and a serjeant remains to mark each flank.——The whole are ordered to take 3 Side Steps to the right, at the word March, in order to disengage.
March. Front.	At the second word March, the whole move on, and each file wheels successively into the center as it arrives at, and beyond the colours.—As soon as each company is in the line from the colours to the flank serjeant, its officer *fronts* it.—When the whole is formed the colours countermarch, and the whole are looking to the colours till otherwise ordered.

In the countermarch from both flanks no part of the battalion is fronted till the whole is on its ground.—In the countermarch from the center, the battalion begins instantly and successively to front by companies, as each is ready and on its ground.

S. 99.   *Countermarch of the Battalion or Line on its Center by Companies, or Subdivisions.*

Fig. 40. A.
When a whole battalion is to countermarch on a central point; although it may be done by files, yet without great care it will be apt to open out: such, or a larger body will best and quickest make such countermarch by the march of columns of companies or subdivisions in front.

One

One or two central subdivisions wheel the half circle upon their center point; or countermarch into the new line, so that the front rank shall occupy the ground which the rear rank did, and the battalion is CAUTIONED to countermarch from its center by subdivisions.

> CAUTION.

One of the wings FACES to the right about: both wings WHEEL inwards by subdivisions: they MARCH along the rear and front of the formed division, and successively *wheel up* into their respective places on each side of those already arranged in the line.

> WING ABOUT FACE.
> SUBDIVISIONS
> INWARDS WHEEL.
> MARCH.
> *Halt, Dress*.
> MARCH.
> *Wheel*.
> *Halt, Dress*.

The subdivisions which wheel up to the rear, successively *Halt, Front, Dress*, when they come to their ground, and the officers who command them, must take care not to pass the rear, but to be at their proper front rank when they *Halt, Front* their subdivisions.

> *Halt, Front*.
> *Dress*.

Should it be intended that the front rank of the directing company or subdivisions should stand on the identical line it occupied before the countermarch, it will be so placed; and in that case after the subdivisions had wheeled inwards, the wing which was to march in rear of it would shift a few paces to the flank, in order to get clear of the rear ranks, and would then be put in march.

When at any time, one flank of a battalion or line, is to be placed at the spot, where the other one stands, it cannot be done in a shorter manner than by prolonging the new line.—If the flanks are to exchange place with each other, the countermarch on the center or on a flank muſt effect it: the ſingle battalion may do it by files; but a line muſt do it by countermarch of diviſions in open column.

COUNTERMARCH IN COLUMN.

S. 100. *When the Battalion Column (or a more conſiderable one) countermarches each diviſion by Files, ſo as to change its Front,*
Fig. 40. B. *and face to its former Rear.*

COUNTERMARCH BY FILES.

RIGHT, FACE.

{ *If the Column ſtands with the Right in Front.*——A CAUTION to countermarch is given.—At the word right FACE the whole face to the right, each company officer will immediately quit the pivot, and place himſelf on the right of his company, and his covering ſerjeant will advance to the ſpot which he has quitted, and face to the right about.

MARCH.

Halt, Front.

Dreſs.

{ At the word MARCH the whole move, the officer wheels ſhort round to the right, and proceeds, followed by his files of men, till he has placed his pivot front rank man cloſe to his ſerjeant who remains immoveable.—Each officer inſtantly gives the words *Halt, Front—Dreſs* to his company, ſo as to have it ſquare and cloſed into the right which is now the pivot flank, and on which the officer now replaces his ſerjeant

jeant who falls back behind the rear rank.—In this manner the column will face to its former rear.

If the column stands with the left in front.—The CAUTION to countermarch is given.—At the word left FACE the whole face to the left, the officer moves to the left of his company, and the serjeant occupies his place, and faces about.—At the word MARCH the officer wheels short to the left and proceeds as before, till he is fixed on the pivot flank, now the left, as the column stands with its right in front.

COUNTERMARCH BY FILES.

LEFT, FACE.

MARCH.

Halt, Front.

Dress.

In the countermarch, the facing is always to that hand which is not the pivot, but which is to become such.

This countermarch of each division separately on its own ground, is an evolution of great utility on many occasions.—It enables a column which has its right in front, and is marching in an alignement, to return along that same line, by becoming a column with its left in front, and to take such new positions in it as circumstances may require, without inverting or altering the proper front of the line.—In many situations of forming from column into line, it becomes a necessary previous operation.

When a column countermarches by divisions each on its own ground, unless the divisions are equal the distances after the countermarch will not be the true wheeling distances, but will be such as are equal to the front of the preceding division, and therefore the true distances must be regained, before the divisions can truly wheel up into line.

S. 101. *When the leading Flank of the Column is changed by the successive March of Divisions from the Rear to the Front.*

Fig. 41. A.

HALT.

LEFT WING TO THE FRONT.

Right, Face.
Quick, March.

Halt, Front.
March.

⎧ If the right is in front, the left to be brought up, and the column to continue to advance.—The whole is ordered to HALT.—At the caution LEFT WING to the front, the officer of the left (the rear) company immediately orders it, *Right Face—Quick March*, till his left flank can freely pass near the right flank of the others.—He then commands *Halt, Front—March* (in ordinary time) close by the right flank of the company then preceding him.

Right, Face.
Quick, March.

Halt, Front.
March.

⎧ The officer commanding that company as soon as the other approaches him, orders, *Right, Face—Quick, March*, behind the now leading one.——*Halt, Front* when he covers—and then *March* when at the due wheeling distance.—All the other companies successively perform the same operation; and when the right company has taken its place in the rear, the whole column is in perfect order.

Fig. 41. B. If before this operation the column should be closed to half or quarter distance, then all the companies may be FACED at the same time, proceed as above directed, and each takes its distance from its preceding one, before it moves on.

This operation is often required in taking up original positions from column of march.—It changes the leading flanks of a battalion or a

more

more confiderable column, and enables it to enter on a line which unforeseen circumſtances require it ſhould prolong.—It permits battalion columns aſſembled at a rendezvous, to march off from whatever flank is moſt advantageous, for each to enter on its line of formation.—It prepares a column which has expected to form by wheels to its left, to be ready to form by wheels to its right, without inverting its order.— In a column compoſed of ſeveral battalions where an inverſion of the battalions within themſelves, but not of the wings is meant to be prevented; then each battalion ſeparately will perform this operation; but if the inverſion of the wings alſo is to be avoided, then the whole column will proceed, as if it was a ſingle battalion.

It muſt be obſerved as a general principle, that the diviſions which advance, come out always on the ſide to which front is to be made, and on which the enemy is placed, becauſe then with the diviſions which are free he can be oppoſed, while the others are moving behind the line.

S. 102. *When the Column changes its Wings, on*
Fig. 42. C. *the Ground on which it then ſtands.*

The left or rear company proceeds as has been already directed: All the others go to the RIGHT about, and MARCH on at the ordinary ſtep towards the place from whence the left moved.	HALT. LEFT WING TO THE FRONT. RIGHT ABOUT FACE. MARCH.

When

Left Face.

Quick, March.

Halt, Front.

March.

{ When the company next it arrives at that place, it receives the order, *Left Face—Quick March* behind the left company, then *Halt, Front,* and *March* when at its due distance.——In this manner all the rest proceed, till the right company when it fronts, finds itself where the left originally stood; only that the whole column is removed to the right a space equal to its front.

S. 103.
Fig. 42. D.

When the Column changes its Wings by the Divisions marching through each other, from Rear to Front.

COMPANIES TO RIGHT AND LEFT OPEN.

MARCH.

HALT.

{ The column standing marched from the right should naturally form to the left, but it is here intended to form to the right.——At the word COMPANIES to the right, and left OPEN—MARCH, all the companies (except the last) do open by the side step, half to each flank, a space sufficient to allow a company to march through in front.

March.

Close inwards.

March.

Halt.

March.

{ The left company does not open, but *marches* on through the others, and as soon as its rear rank arrives at the front rank of the one next it, that company closes by the side step, *marches* and follows at its due distance: In this manner they succeed each other, till the column is formed as marched off from the left.

But

But if the ground of the column is not to be changed after opening out; the laſt company *moves* on, after the others having FACED about, and MARCHED, have arrived at its ground; each there ſucceſſively *faces* inwards and joins, then *fronts* and *marches* on till the word HALT is given, when the flanks are changed, and that the left company is exactly on the ground where the right ſtood.—The leading company muſt take ſhort paces to allow for the various operations of the following one.

> RIGHT ABOUT FACE.
> MARCH.
> Inwards Turn.
> Front.
> March.
> HALT.

The above method of countermarch is more calculated for a parade than for the general movements of the battalion.

GENERAL CHANGES OF POSITION OF THE BATTALION.

CHANGES of POSITION of the battalion or line from one diſtant ſituation to another are made either in *Line* or by the *Echellon* march of diviſions; or by the movements of the column, eſpecially of the *Open Column*.

Changes of poſition in OPEN COLUMN, are movements of previous diſpoſition, made from one diſtant ſituation to another, and not liable to the interruption of an enemy.——Where circumſtances allow,

> By the movements in open column.

original

original or new positions are in this manner, easiest and soonest taken up.

Changes of position of the Battalion or Line already formed, when made in one or more Open Columns, may be divided into 4 Parts.

1st. The line wheels the quarter circle by platoons or such other divisions as are ordered to either hand, so as to be ready to divide into one or more columns.

2d. The column or columns file by divisions or march in front, as is necessary and ordered to arrive at their position in the new direction.

3d. The divisions again form in a general open column, perpendicular to the new direction.

4th. When the divisions of each battalion are thus arrived at their ground; halted, and adjusted, the line is formed by their wheeling up,——and thus battalion after battalion; each forming when its adjoining one has 3 or 4 of its divisions standing in column on the line.

By the movements in Echellon.

The ECHELLON changes of position are the safest that can be employed, in the presence of and near to an enemy, they are almost equal in security to the march of the line in front, or to an uniform wheel of the line, but which is not to be attempted; they can be used in the most critical situations, where the filings, and movements of the open column could not be risked; they are more particularly used when the enemy's flank is to be taken by throwing the body forward, or when
one's

one's own is to be covered by throwing it backward.——The advantages attending them are; the preserving a general front during the march, and allowing sufficient freedom of movement, which in such situation is indispensible; they enable to change position on any division of the line, either on a fixed or moving point; and at any instant the movement can be stopped, the line formed, and a sudden attack repulsed.——The echellon changes require the ground to be nearly of such a nature as a full line could advance in; and any of its divisions that meet with obstacles in their march will pass them in the same manner as they do in line, by filing or doubling, and without interrupting the progress of the others.

Changes of position of the Battalion or Line made by the Echellon march of Platoons, consists of 3 parts.

1st. The platoons wheel forward a certain number of paces towards the hand to which they are to change position, and so as that each thereby stands perpendicular to its future line of march.

2d. Each platoon marches on directly in front, to its proper point in the new line.

3. Each platoon successively on its leading flank arriving at the platoon preceding it, (which is already halted in the line) dresses up and forms truly in that line.

———

Each change of position of the battalion, or line, may be considered as a general wheel of the whole made on a POINT, either IN, BEFORE, or BEHIND, the old line.——The battalion or line therefore breaks, to

which

which ever hand, and to which ever divifion it is to manœuvre to or be led by: When to a flank; generally to that which is neareft to, and is fuft to enter any part of the new pofition: When a central divifion determines its movement, it breaks to right or left inwards and faces fuch divifion, which makes its change of fituation on its own ground. ——When this POINT is IN the old line; it muft neceffarily be within the battalion when fingle, or within a certain named battalion of a line: Such battalion therefore will have to perform the change on a *fixed* point within itfelf, viz. on fuch divifion flank or central, as is already refted at that point; by making its other divifions either by *filing*, or *diagonal* marching, enter into the line: But all the other battalions will have the double operation of moving up to the new line, and then forming upon it.—When this POINT is BEFORE or BEHIND the old line, every battalion whatever fingle or connected will have this double operation to perform.

Fig. 43.

Fig. 44.

---

§. 104. *Changes of Pofition of a Battalion.*

The battalion formed in line, changes to a new pofition either on a fixed point within itfelf, or on a diftant point, which marks one of its future flanks, or where one of its central divifions is to be placed.

When on a fixed Point either Flank, or Central.

1ft. By the echellon march of divifions either to front or rear,

Fig. 46. A. rear, which move on and line with the placed or fixed one, when it halts on its ground. (S. 159.)

Fig. 46. B. 2d. By breaking into open column so as to face the fixed point.—Filing divisions to front or rear, into the new direction, and wheeling up into line, when the column is prepared. (S. 120. 121.)

When on a distant Point, and that the Whole are moveable.

Fig. 57. A. No. 1. By the echellon wheel of divisions, and the subsequent march of the whole, till the one nearest to the new line arrives in it, and that the others move on, and form to it. (S. 162.)

Fig. 57. B. No. 2. By the breaking into open column to the one or other flank, and the immediate filing of all the divisions from the old line to the new one. (S. 123.)

Fig. 57. C. No. 3. By the march of the battalion column to the point where its head is to rest, and then facing, and filing its divisions into the new line. (S. 124.)

Fig. 57. F. No. 4. By the march of the battalion column, and its wheeling into the new line, at the point where its rear is to rest. (S. 125.)

No. 5. By the march of the battalion column, and its
Fig. 57. H. wheeling into the new line at a point where one of its central divisions is to reft. (S. 126.)

Befides the above which are the moft general modes by which changes of pofition fhould be effected by the battalion, the open column on entering its ground, may alfo occafionally be required to form in line in the following manner.

No. 6. When the column having arrived perpendicularly or obliquely behind the line at the point where its HEAD is
Fig. 52. 78. to reft, is there halted.——The leading divifion may be placed on the line, and each other divifion be ordered to make fuch a degree of wheel backwards, as will enable it to march on in front, perpendicular to its proper point in the new line, where each fucceffively arrives and forms.—This is a movement in column, and formation in echellon. (S. 127.)

No. 7. The column arriving in the direction of the line, or in any direction oblique, or perpendicular behind the new line and at the point where its HEAD is to reft, but
Fig. 57. D. which its rear is to pafs.——May form by the wheel of the
Fig. 55. leading divifion into the new line, and the fucceffive march of the other divifions behind it and behind each other, till they arrive at their feveral points of wheeling up. (S. 128.)

No. 8. The column marching perpendicularly up to the line, and to the point where its HEAD is to reft, and being at leaft a diftance equal to the length of the column from

fuch

Fig. 57. F. such point.———The leading division proceeds at a half pace only; the others oblique from the column, successively move up to the leading division, and the front being thus gradually encreased the whole battalion arrives at the same time on the line of formation. (S. 29.)

The column arriving behind any part of its ground may also move up to close column, and form by its deployments on the front, the rear, or on a central division.

———

S. 105. A battalion broken into, and marching in open column, must arrive at, and enter on the ground on which it is to form in line, either—In the DIRECTION of that line: PERPENDICULAR to that line: or in a direction more or less OBLIQUE, and betwixt the other two.

If where its HEAD is to rest.—The leading division will wheel up into line, and the others march on behind it, and successively wheel up as in No. 7.

If where its REAR is to rest.—It marches with its pivot flank, and at just distances along the line, till the rear platoon is at its point, the whole then halt, and wheel up into line as in No. 4.

If the Column is marching in the direction of the Line, it will either enter where its head is to rest, or where its rear is to rest.

Fig. 58. A.

If

if the Column enters perpendicular, or oblique to the new line, it will enter either, where its head is to rest, where its rear is to rest, or at some intermediate point where a central division is to rest.

Fig. 58. B. C.

If where its HEAD is to rest.—The formation may be made as in No. 3.

If where its REAR is to rest.—The formation may be made as in No. 4.

If at an intermediate point where a central division is to rest.—The formation may be made as in No. 5.

Relative situation of old and new positions.

All new positions, that a battalion or line can take with respect to the old one, are—PARALLEL, or nearly so to the old line.—INTERSECTING by themselves or their prolongation some part of the old line or its prolongation.

Parallel.

Fig 44. 63.

New PARALLEL positions being necessarily to the front or rear of the old one, the battalion will according to circumstances take them up by the *Echellon* march, the *filing* of divisions, or the *Movement* in open column, and its subsequent formation in line.

Intersecting.

Fig 43. 63.

New INTERSECTING positions, which themselves cut the battalion, will be taken up by the *Echellon* march, or by the filing of divisions.— All other new positions which themselves or their prolongation, intersect the old line, or its prolongation, will in general be taken up by the *March* in open column, and its subsequent formations when it arrives at the line; some such positions will however allow of, and require being made by the *Echellon* march, or by the *filing* of divisions.—

In

In general the battalion will break to the hand, which is neareſt to the new poſition, be conducted to its neareſt point in the new line, and form on it as directed.

In changes of poſition by the open column, the whole battalion (as a general rule) is directed to wheel the whole quarter circle into open column, although it may often ſeem an unneceſſary operation, and that diſengaging the heads of diviſions would anſwer the ſame purpoſe, where the change is to be performed by *filing*.—Yet is the above general mode to be obſerved, becauſe it is a poſitive and defined ſituation, from which every change can proceed, whereas all other modes are liable to uncertainty and miſtake, and the apparent going over a little unneceſſary ground is a matter of no moment in point of time, and begets perfect preciſion, and correctneſs of execution.

<small>When the battalion breaks into open column to make a change of poſition.</small>

This rule which is univerſal for all the following diviſions of a column, may in ſome ſituations be diſpenſed with as to a leading diviſion which often has to wheel up again over the ſame ground, when the column is put in motion towards its new poſition: A previous and ſeparate CAUTION from the chief of the battalion may therefore when it is ſeen neceſſary, prevent this extra movement to the leading diviſion and give it a more favourable ſituation in the direction in which it is to proceed.—As in the caſe of a battalion marching off by column of diviſions from a flank to the front.

If the Wheel is made backwards. The flank diviſion may ſtand faſt, till the wheel is made, and when the reſt of the column has marched up up to it, it then receives the word March from its own leader, and proceeds.

If the Wheel is made forward. At the first word March, the flank division moves on a space equal to its own front and halts, it is then ready to proceed when the whole is put in motion.

Should the battalion march off by column of divisions from a flank to the rear: Whether the divisions *wheel backwards or forwards* the flank division wheels with the others, and from that situation, that division will again wheel to the rear, when the column is put in motion.

In central changes of position, the battalion or line, breaks into open column, facing to the named division.

Fig. 45.

In all central changes of position on a point within a battalion or line, and which are made by the movements of the open column: The battalion or line breaks backwards into two open columns facing each other and the given point, so that the one has its right in front, and the other its left.—From this situation by the filing, or by the march of divisions, its component parts move to their new position, and the division which faces the given one having there taken a double wheeling distance, the divisions wheel up into line.

The advantages of making central changes, by breaking inwards, so as the whole stand faced to the named division, in two columns, are—That the universal rule of all bodies breaking, dressing, forming to whatever point they are led by and manœuvre to, is observed.—That the taking of distances in the new column are all from the front and none from the rear, which last is a matter of difficulty and delay.—

That

That the battalions of the wing which is thrown forward, advance from their inward flank and in the shortest line, to where that flank is again to be placed, at this point they begin to form, and the formation is made by quick filing of divisions into the new column, where the exact covering of pivots and taking of distances is instantly and easily ascertained.—That the parts of the line on each side of the central division work exactly in the same manner, and form in line by one and the same method.—That the breaking inwards of the line or the countermarch of such part of the column as is before the central division gives these advantages, nor is the countermarch the affair of a moment.

In *central Changes* was the whole of a line to break to one hand, or part of a column not to countermarch.—Although such part as was behind the central division, and thrown backwards would take its distances from the front and might proceed exactly as above: Yet such part of each as was before the central division would be obliged to take its distances and covering of pivots in the new column from the rear as the whole line would be broken the same way as the named division. —This though it may not seem difficult when such part consists only of the few divisions of a single battalion, will when it is composed of several battalions in addition, be found no easy matter to accomplish with precision.—In such case whatever divisions of the central battalion were arranged before the named division, would file from their advanced flanks and place themselves in the above manner in the new line. The entire battalions which were before the named division would march in separate columns of divisions, each from its head or outward flank, and enter (by wheeling) the new line, at the point where its rear or inward flank was to be placed, it must then prolong the line and be halted the instant the rear arrived at the point where the head entered.—This operation would not be found eafy, be flower, and attended with more uncertainty, than the other method by which the diftances are fo readily, and exactly taken from the front, and where the fame mode of execution is followed by both flanks of the line —Although battalions and lines fhould be prepared to change their pofition in this manner if fo required; yet the other method is to be confidered as the general one, and practifed accordingly.

WHEEL OF THE BATTALION FROM LINE INTO OPEN COLUMN.—CHANGE OF DIRECTION OF THE MARCH. —WHEEL AND ENTRY ON AN ALIGNEMENT.— MARCH.—HALT —AND WHEEL UP INTO LINE.

S, 106. *When the Battalion halted in Line, Wheels forward by Companies into Open Column, the Right in Front.*

COMPANIES, RIGHT WHEEL.
{ At the CAUTION Companies Right Wheel.—The officers ftep out nimbly, and place themfelves one pace before the center of the companies facing to the front; at the fame time the right hand man of the front rank of each company faces carefully on his left heel to the right and becomes the pivot, on which each company is to wheel. The covering ferjeant

serjeant of the right company also runs out and places him- | Fig 59. A.
self at the point (a) where the wheeling flank of that company is to *Halt* at the finishing of the wheel.—The covering
serjeants of the whole fall back two paces.—The supernumerary rank closes up within two paces of the rear rank and
the divisions of drummers, &c. enter into it, behind the
respective companies which they cover, or are divided behind
their several companies.

At the word MARCH each company steps off quick, turn- | Q. MARCH.
ing eyes (and not before) to the wheeling man, and carefully observing the general wheeling directions.—The left,
or wheeling man takes his firm lengthy step of 33 inches,
neither opening from nor pressing on his own pivot, and
turning his eyes towards that pivot.—The officer during the
wheel turns towards his men and inclines to his new pivot,
or left flank; and standing faced to it with a glance of the
eye he sees when the quarter circle is compleated, and each
gives his word *Halt, Dress* at the instant that the flank man | *Halt, Dress.*
is taking the last step which finishes his wheel perfectly
square.—The officer immediately corrects any dressing that
the company may require within itself, instantly places himself on the pivot flank, and his serjeant covers the second file
from that flank.—Both colours wheel up into column, and
at all times remain behind the third file from the pivot flank
of the leading center company, whether the company is halted or in motion.

S. 107. *When the Battalion halted in Line, Wheels forward by Companies into Open Column, the Left in Front.*

COMPANIES, LEFT WHEEL.
Q. MARCH
Halt, Dress.
{ The same operation takes place as in wheeling to the right, with these variations; that the left hand men of companies face before the wheel begins, and the left covering serjeant marks the ground for the flank of the leading company.

After the battalion has in this manner wheeled forward into column, it will often happen that from the inequality of divisions, different sizes of men, &c. &c. the pivots do not exactly cover; yet in this situation are they to remain and to understand it as an invariable rule, that they are never to shift in order to cover, but by the express direction of the commanding officer, who will correct the pivots, if his intention is to pursue a straight line in order to form; but if the continuation of a march is the object, he will allow them gradually to get into its direction after they are put in movement.——But the certain remedy for the above inconvenience is, that on all occasions of wheeling into open column from line, the wheels shall be made BACKWARD instead of forward.

S. 108. *When the Battalion halted in Line, Wheels backwards into open Column, the Right in Front.*

COMPANIES ON THE LEFT BACKWARDS WHEEL.
{ At the CAUTION, companies on the left backwards wheel; the officers step out nimbly and place themselves before the center of their companies, facing to the front, at the same time

time the left hand man of the front rank of each company *faces* carefully on his left heel to the right, and becomes the pivot, on which each company is to wheel.—The covering ferjeant of the right company alfo runs back, and places himfelf at the point (S) where the wheeling flank of that company is to halt at the finifhing of the wheel.—The covering ferjeants of the whole fall back two paces.—The fupernumerary rank clofes up within two paces of the rear rank, and the divifions of drummers, &c. enter into it, behind the refpective companies which they cover, or are divided behind their feveral companies. } Fig. 59. B.

At the word MARCH, each company fteps back quick and follows exactly the fame directions that have been given in the cafe of wheeling forward. } Q. MARCH.
Halt, Drefs.

S. 109. *When the Battalion halted in Line, Wheels backward by Companies into Open Column, the Left in Front.*

The fame operation takes place as when the right is in front, except that the right hand men of companies are the facers, and the left ferjeant marks the ground for the flank of the leading company. } COMPANIES ON THE RIGHT, BACKWARDS WHEEL.
Q. MARCH.
Halt, Drefs.

S. 110.

S. 110. *If the Battalion is at once to break into Column of Sub-divisions or Sections.*

CAUTION.

Q. MARCH.

Halt, Dress.

The pivot men of each *face* and their divisions wheel into column at the general word MARCH; the company officers (only) give the word *Halt Dress* which fuffices for the parts of each company.—When the wheel is compleated and not before, the leaders who are to conduct the pivot flank of the second fubdivifion, or of the second or other sections place themselves there.—The officer is on the pivot flank of the leading fubdivifion, or section; his covering serjeant on the flank of the second fubdivifion, or second section; and an officer or non-commiffioned officer from the rear on the flank of the laft section, after wheeling into column. (S. 47. 48.)

S. 111. *When the Open Column is put in March in the Prolongation of the Line.*

MARCH.

The battalion ftanding in open column with the pivot flanks of its divifions on the line, and advanced points being afcertained, moves forwards at the word MARCH from its commanding officer. (S. 115.)

Whenever the battalion wheels into open column in order to prolong the line on which it was formed, and that no diftant point in that prolongation is previoufly given.—The

serjeant

serjeant of the leading company will advance 15 or 20 paces, and place himself in the line of the pivot flanks, and the leading officer will thereby (taking a line over his head) be enabled to ascertain the direction in which he is to move.

S. 112. *When the Open Column with the Right in Front changes Direction to the Left, on a moveable Pivot.*

Right Shoulder forward.
Forward.
} As explained in S. 22. 52.

S. 113. *When the Open Column with the Right in Front changes Direction to the Right, on a moveable Pivot.*

Left Shoulder forward.
Forward.
} As explained in S. 22. 52.

S. 114. *When the Open Column advancing with the Right in Front, Wheels on a fixed Point into a new Alignement.*

The alignement is entered by the leading division wheeling either to right or left.—In either case the left or pivot flank officers of the companies

[96]

panies muſt be placed on it: in the firſt inſtance behind it, and in the ſecond before it.—In both caſes the line is afterwards formed by wheels of companies to the left: in the firſt inſtance the line will front the ſame way as the column; in the ſecond, it will front to the rear of the column.

S. 115. *When the Open Column advancing with the Right in Front, Wheels to the Right, on a halted Pivot into a new Alignement, and marches in it.*

Fig. 60. C.
Fig. 52.

Right. Wheel.

Halt, Dreſs.

March.

The alignement being determined by given objects, and the point (c) of entry marked; the leading officer who has marched his left flank on that point, when he arrives at a diſtance equal to the front of his company from it, orders *Right, Wheel!* and the quick wheel is made, ſo, that on the concluſion of it at the word *Halt, Dreſs*, he himſelf ſhall be ſtanding on the new alignement on the flank of his company ready to give the word *March* as ſoon as the ſucceeding company has arrived at the wheeling point.

After this he moves on without looking behind, regarding his diviſion, or allowing any thing to take off his attention, and at the eſtabliſhed ordinary pace towards the diſtant points (x, a) ſo that his ſhoulder ſhall juſt graze the head of any mounted officer's horſe poſted at an intermediate point (or the breaſt of any man on foot placed for the ſame purpoſe) and which he invariably preſerves in a ſtraight line with the given object.——This rule all the following officers

muſt

must observe at the same time that they maintain their exact distance from the company preceding. And should any of the companies deviate to either hand, those that succeed them must rectify the fault, and exactly touch the point where the adjutant is placed.

The principal attention of the leading officer must be, never to change the time or length of step, otherwise a stop must happen in a considerable column and the soldiers will afterwards be obliged to run. He must move in one constant position with his front rank perpendicular to the line on which he marches.—The same directions regard the other officers who conduct companies, and who in addition must correctly observe, that at the word *March* given to the preceding company, the following one is ordered *Right, Wheel*. In this they will exactly agree if the officers preserve their due distances, and make their wheels at a redoubled pace, and also, that all the companies wheel at the identical point where the leading one wheeled; therefore all the companies must march straight up to the point where the first rank of the preceding one commenced its wheel.—The attention of pivot officers marching in the alignement, have been already described in the open column.

To insure the more correct march and halt of the pivot flanks in the alignement.—The commander of the battalion or column may occasionally go forward to an advanced adjutant, and being himself truly placed, may look back to the point of wheeling or entry into the alignement, or to any other fixed object that is in it.—He can then see if the rear flanks of the column keep the true line, or deviate from it, and may correct them by signal; or by sending back an adjutant to take his position in the true line, and to whose direction they are immediately to conform.

In this manner also can the leader if necessary correct the pivot flanks after a halt, when there is a rear point of view sufficiently marked.—If that is not the case, he may go towards the rear of the column, line the flank of the 5th or 6th company, on that of his leading company, and a front point of march, he will then return to the first company, and on the flanks of that and the 5th correct the rest of the pivots.

S. 116. *When the Open Column advancing with the Right in Front—Wheels to the Left on a fixed Point, into a new Alignement, and marches in it.*

Fig. 60. B.

Left Wheel.
Halt, Dress.
March.
{ The leading company begins its wheel to the left on the alignement itself when its pivot flank officer arrives at the point of wheeling, instead of (as in the preceding section) beginning at the distance of a company short of that point. (S. 51.) }

Whatever has been said respecting a battalion broken from the right, takes place in one broken from the left: the only difference is, that the flanks are now changed; that the left company does what before was done by the right; and that the right flank officers are placed on the alignement instead of the left.

§. 117. *When the Open Column advancing with the Right in Front, and composed of Divisions of unequal Strength, Wheels to the Right on a halted Point, into a new Alignement.*

Fig. 62.

The pivot or left flank continues to direct 'till the leading division arrives in its full front behind its proper wheeling ground and at a due distance from it.—The word *Right Wheel* being then given, the reverse or right flank (c) of that division stops, and the general pivot one compleats the wheel, so that at the next words *Halt, Dress, March*, the conducting officer may be exactly placed on the new line of direction; they thus succeed each other, observing that a stronger division (a) wheels short of the ground of its preceding weaker one (b) by the space of as many files as it exceeds that preceding one; and a weaker division overpasses the ground of its preceding one, by the extent of as many files as it is deficient: in both cases after the wheel, the divisions will have retained the same relative situations as before its commencement and the left pivot flanks will still cover.

Right Wheel.

Halt, Dress.

March.

§. 118. *When the Open Column——Halts——Wheels up into Line, and Dresses.*

Fig. 60. 61.

If in the manner already directed, the several companies of one or more battalions have entered the alignement and marched with their pivot flanks along it, covering each other at their due diftances for which company-officers are anfwerable, there can be nothing eafier than to form well in line.

HALT. { Whenever therefore the head or the rear divifion arrives at the given point where it is to reft in line; the commander of the battalion gives the word HALT.—No one moves after the delivery of this word, not even a half pace, but the foot which is then off the ground, finifhes its proper ftep, and the other is brought up to it.—If that was not done, and that one company fhould ftop while another was permitted to make one or two paces, thofe behind would be obliged to fhift anew, and much confufion would arife from officers being deficient in one great principle of their bufinefs the preferving of proper diftances.—The inftant the HALT is ordered, the commanding officer from the head divifion of each battalion (he taking care that he is himfelf placed in the true line) makes any fmall correction on a rear point in that line that the pivots may require, although no fuch correction ought to be neceffary.

WHEEL UP INTO LINE.

Fig. 49.

{ The CAUTION is then given, companies wheel up into line; on which the pivot men of the front ranks face perfectly fquare into the new line; the company officers move brifkly out and place themfelves one pace before the center of each, their covering ferjeants move to the right of the front rank of the companies if the wheel is to be to the left, or otherwife behind the pivot file if the wheel is to be to the right, and an under officer of the leading company of the battalion

battalion runs up, places himself square in the new line, and marks the point (s), at which the wheeling flank of that company is to arrive and be halted.

At the word MARCH, eyes are turned (and not before) to the wheeling hand, the whole step off in quick time, the wheeling man lengthening his step to 33 inches, and every other man diminishing his, as he is nearer to the standing flank.—The officers during the wheel turn round to face their men, incline towards the pivot of the preceding company, and as each perceives his wheeling man make the step which brings him up to that pivot; he gives the word *Halt, Dress*, strong and firm to his company, which halts with eyes still turned to the wheeling flank, and each officer being then placed before the preceding pivot to which his men are then looking, from thence corrects the interior of his company, upon that pivot, his own pivot, and the general line of the other pivots.—This being quickly and instantaneously done, the officer immediately takes his post on the right of his company, which has been preserved for him by his serjeant.

Q. MARCH.

Halt, Dress.

In this manner dressing is made, and eyes are turned always to the point where the head of the column halted; to the right when the wheels are made to the left, and to the left when the wheels are made to the right; and if any future correction of the line is made by a field officer, it will be from the fixed point where the head of the column rested.

As there are so many determined points given, it becomes easy to dress correctly a platoon or battalion after wheeling up, if due care is taken that the pivot man do on no account move up, or fall back, whatever directions may be then giving by the company officers for compleating the dressing.—If a defect exists it must proceed from the other men not having lined with those fixed points; the internal correction of companies must therefore be made, but the original pivot men remain immoveable, until a general correction of dressing the battalion is made by a field officer if necessary.—The officer of the third company for example, if the wheel has been made to the left, has only to consider the left file leader of the second company close to whom he stands as the point of Appui, and his own left flank man as the point to dress upon, there will then be nothing easier than to dress the other men of his company upon these; but he will still more exactly do it, if he places himself 2 or 3 files on the other side of the pivot man of the second company, and from thence corrects his own.—If all officers are in this alert and skilful, and that soldiers are accustomed to dress themselves, a battalion will be instantly formed, nor will the commander have any thing to rectify.

When the column has broken to the LEFT; all that has been before said takes place; and is in the same manner executed.—Only the right flank man does what has been directed for the left; he fronts when the platoon begins to wheel up; and the point d'Appui being now on the left, the dressing must from thence be regulated, consequently the soldiers look to the left.

It is to be observed that when at any time after forming in line, there shall be a false distance between either of the flank divisions and the battalion,

talion, the officer of such division, without waiting for directions, may immediately by the closing step, join his division to the battalion: but no other division of a battalion is in such case ever to move, without orders from the commanding officer.

When the battalion has formed in line, and that there are several false openings betwixt divisions, they may be remedied by the closing step on the order of the commanding officer to CLOSE to any named division, the others halting successively by word from their several leaders. S. 43. 79. and in the same manner may the crowding of files in a battalion, or parade be remedied, by closing from the point of crouding, and halting when sufficiently loosened.

S. 119. *When the Open Column which is to wheel into Line is composed of Subdivisions, or Sections, and not of Companies.*

At the word WHEEL, the company officer alone moves into the front, and the pivot leaders of the other subdivision or sections go to the point they would be at, if the column was a column of companies: The pivot man of each body in the column faces.—At the word MARCH, the whole wheel. —And the company officer gives the word *Halt, Dress*, to the whole company. (S. 50.)

{ WHEEL INTO LINE.

Q. MARCH.

Halt, Dress.

The line of the march of the open column will always be about a pace before the line on which the troops form; because the one is the direction

tion preserved by the officers in marching, and the other being that on which the flank men halt and the companies wheel up into line, is distant of course from the first the breadth of a file, which leaving the advanced points distinct, affords a great advantage in the formation and correction of the line.—Although the officers halt in the alignement itself, yet it is impossible to allow them to remain immoveable as points of forming for their divisions, because the dressing of those divisions depends on them, and that they must occupy their proper places when in line.—The flank files of men are therefore the pivots of divisions in wheeling up into line, although the officers are the pivots during the march, and an attempt to form the line on the points of march themselves would derange the pivot files of men, and cause disorder.

A commander must be careful that he himself is in the alignement, whenever he dresses his battalion, or corrects the flanks of his divisions. —In order to direct well, he must place himself on the line, and on the adjutants who are in it, and give his horse such a direction, as the divisions should touch in marching.

Changes of Position of the Battalion from Line, by Movements of the Open Column.

Changes of position are made either on a FIXED point within the battalion, or on a DISTANT point without it.

On

On a Fixed Point.

S. 120. *If the battalion is to change position to the Front, on the right halted Platoon, by throwing forward the whole Left, and by the filing*

Fig. 47. B. *of Platoons.*

The right flank (c) is the fixed point on which the change is made, and is in the intersection of both lines, the commander immediately ad libitum, places another point (b) 20 or 30 paces beyond that flank, these two determine the direction of the new line and face to it.—The right platoon is *wheeled* forward to the right and placed in that direction, and is then immediately *wheeled* backward on the left, till it stands with its pivot (a) perpendicular to that direction, and on which its officer posts himself.

The rest of the battalion is then wheeled backward on the left, by platoons and stands in open column.—At the word left FACE, the whole (except the fixed platoon) face. } BY COMPANIES ON THE LEFT BACKWARDS WHEEL. LEFT FACE.

At the word Q. MARCH the several officers lead their files towards the points in the new line where the pivot flanks of their platoons ought to be placed, and the better to ascertain those points, the covering serjeant of each platoon will successively (as it approaches within 20 or 30 paces of the new line) run up and place himself upon it at the proper distance of his platoon, facing to the head of the column, and covering } Q. MARCH.

vering exactly those that have taken their places therein: The pivot flank officer (a.) of the front platoon, and the advanced officer or serjeant (b.) before mentioned are the original points on which the first serjeants that come up arrange themselves, and thereby become additional points for the others.

Halt, Front.
Dress.

The serjeant thus placed (being on the spot which the officer is afterwards to occupy; each officer comes up in his own person immediately before the serjeant, *Halts fronts* his platoon, *Dresses* it quickly by closing his flank front rank man to his serjeant, and placing it perpendicular to the new line.—The officer takes the place of his serjeant, and the whole being steady, and pivots corrected by the commanding officer as they arrive upon the line, every one is in a situation to wheel up and form.——Should no serjeant be previously advanced to give the pivot point, the officer must at once conduct the head of his file to it.

WHEEL UP AND FORM.

S. 121.

Fig. 47. C.

If the Battalion is to change position to the Rear, on the right halted Platoon, by throwing back the whole Left, and by the filing of Platoons.

ON THE LEFT BACKWARDS WHEEL.

The direction of the line being ascertained in the before mentioned manner, the right platoon is wheeled back on the right into the line, and then backwards on the left, 'till its left or pivot flank (a.) stands perpendicular to the new line.

line.—The battalion will break into open column on the left ⎫ RIGHT FACE.
backwards.—The platoons will face to the right, and the ⎬
officers place themselves to lead. ⎭

At the word MARCH, the whole will lead to the rear, ⎫ Q. MARCH.
and the covering serjeants will successively as before take up
their pivot points on the new line.—The officer conducting
each platoon when he arrives at his serjeant will stop directly before him, allow his platoon to move on behind the serjeant 'till the rear file comes close to, but beyond him; *Halt front.*
The officer will then *Halt, front—Dress* his platoon to the *Dress.*
left, perpendicular to the new direction, and with his front
rank closed into the serjeant.—He will himself take the place WHEEL UP,
of the serjeant, and remain steady on the pivot flank, ready AND FORM.
to wheel into line. ⎭

When the position is changed to the left by throwing the whole right, either backward or forward,—it then follows that the battalion *breaks* on the *right backwards*, that the rights become the pivot flanks, and that the same general circumstances of facing, filing, arranging ser-jeants on the pivot flanks, &c. still take place by the substitution of the commands, right for left, and left for right.

S. 122. *If the Battalion is to change position on a central halted Platoon, by the filing of Pla-toons, and that the Right is thrown forward*
Fig. 48. *and the Left backward.*

One

<table>
<tr><td>Fig. 48.

ON THE RIGHT AND
LEFT BACKWARDS
WHEEL.

RIGHT, FACE.</td><td>One flank of the central platoon is confidered as the point (a.) of interfection, another point (o.) taken ad libitum, determines the direction of the new line.—The given platoon is firft wheeled into it, and then wheeled back till it ftands perpendicular to it; and the covering ferjeant from each of the adjoining platoons runs out and marks where their future pivots (c. b.) are to be placed.—The other platoons wheel backward, fo as that they all ftand faced to the given one.—The whole (except the given platoon) FACE to the right; viz. thofe that are to move towards the front to the front; thofe that are to move towards the rear, to the rear.</td></tr>
<tr><td>Q. MARCH.

Halt, Front.
Drefs.

WHEEL UP,
AND FORM.</td><td>They then MARCH, and the ferjeants giving ground in the line of the pivots which is determined by the three already placed therein, they arrange themfelves in two columns, before and behind the placed platoon, towards which the whole ftill face.——The platoon (b.) which immediately faces to and is next the placed one, muft take care to form with a diftance equal to its own front, and that of the placed one; all the others are at their juft wheeling diftances in column.—From this fituation the line is formed by a wheel to the proper front.</td></tr>
</table>

If the right is to be thrown back and the left forward, the only alteration from the above is that the platoons would FACE to their left, and FILE from their left inftead of their right.

On a distant Point.

S. 123. *When the Battalion is to change to a distant Position either to its Front or Rear, by the filing of all its Platoons, and that this Position is either Parallel or Oblique to the one it quits.*

Fig. 57. B.

The battalion breaks into open column of platoons, to which ever hand the new position outflanks the old one, for to that hand will the whole have to incline during the march; and if it does not sensibly outflank, then the battalion will break to the hand next to the point of intersection of the two lines, for that hand is nearest to, and will in general be the first to enter any part of the new position.

} BY PLATOONS, ON THE ——— BACKWARDS WHEEL.

Q. MARCH.

Halt, Dress.

The battalion standing in open column is ordered to FACE. ——The leader of the second platoon has then a direction given him which crosses the new line at the point (o.) as near as can be judged where the flank of that platoon is to be placed.—The whole are then put in motion.—The leader of the second platoon marches in his given direction at a steady pace; the commander of the battalion remains with the head platoon (c.) and by making it insensibly advance, or keep back, regulates the heads of all the others during the march, as they endeavour to place themselves nearly in the prolonged line of the heads of the two leading platoons,

TO THE — FACE.

Fig. 49.

Q. MARCH.

but

[110]

Halt, Front. Drefs.

but at any rate they are not to be before them; and when thofe two platoons *Halt* their pivots in the line, the others without hurrying arrive fucceffively in the new direction, and ftand in open column at their juft wheeling diftances.—When the head of the column is within 30 or 40 paces of the new line, (its direction being already prepared,) the ferjeants run out and mark the pivot flanks of their feveral platoons.

In this manner the commander who is himfelf with, and conducts the two leading platoons moves them in the direction that beft anfwers his views, and at once takes up any pofition and to any front that is neceffary.—As circumftances change his intentions, he may at every inftant vary, and direct them upon new points of march; the rear of the column always conforming (without the neceffity of fending particular orders) to whatever alterations of direction the head may take; and the commander conducting that head fo as to enable the rear to comply with its movements without hurry.

As the lines of march in filing will feldom be perpendicular to the new line; the leaders of platoons will take care that their laft 12 or 15 paces in approaching their ferjeants, fhall be made in a direction perpendicular to the new line, fo that their platoons may *Halt, Front* juftly; without any neceffity of fhifting their rear files.

During the tranfition from one pofition to another the wheeling diftances fhould be nearly preferved; but at any rate great care muft be taken that they are correct, juft before entering the new line.

When

When the platoons in this manner gain a new position by filing; they always *File* from the flank which is nearest to that position, and place their pivot flanks upon it.——If the pivot happens to be the leading flank, the conducting officer *Halts, fronts* his platoon when he touches the new position, which is marked by his serjeant.—But if the pivot is the following flank, the officer who leads, stops in his own person when he arrives at the new position, marked also by his serjeant, and makes his platoon go beyond it and behind the serjeant, till his pivot man arrives in it, He then *Halts, fronts* the whole platoon.—Conducting officers must therefore recollect that it is always the pivot flanks which are halted in the new position, and that on them the platoons wheel up into line.—In general when the platoons file to the front the pivot flanks lead and arrive first in the new line: When they file to the rear, the pivots follow and arrive last in the line.

Changes of position are thus made in an accurate and expeditious manner by one or two battalions; but an extensive line would be too much broken if thrown into so many small files, nor could it in open ground without the greatest attention to distances risk such an operation, if there was any possibility of an enemy interrupting its completion.—This mode applies in many situations among trees, and where the ground is much impeded with bushes or obstacles which prevent marching on a platoon or a larger front.

When the new line c. outflanks towards the point of intersection, then the battalion breaking to that hand, will have its head (a) nearer to the new line than its rear.——When the new line B outflanks from the point of intersection, then the battalion breaking from that point will have its head (b) farther from the new line than its rear; but in

Fig. 50.

this

this cafe the platoons muft be fo directed during the march by making a kind of gradual wheel forward upon the rear, that the head (b) fhall enter the new line before the rear arrives upon it.

S. 124. *When the Battalion changes Pofition by breaking into Open Column, marching up in Column to the Point where its Head is to remain, and entering the Line by filing its Platoons.*

Fig. 57. C. A.

Fig 51.

HALT.

FACE.

Q. MARCH.

Halt, Front. Drefs.

The pivot flank of the column being directed on the adjutant (c) who marks the flank point in the new line, will HALT when arrived within a few paces of him; a point of direction (d) beyond the adjutant is alfo immediately afcertained.—The word FACE (to the right or left as is neceffary to conduct into the new line) is then given and executed by all the platoons, and the ferjeants begin to run out to mark their pivot points.—— At the word MARCH the whole move in file; the head platoon places its pivot flank at a wheeling diftance from the adjutant, and every other one in the manner before directed arrange themfelves behind the head one, and behind each other; their flanks being corrected by the commanding officer they are then ready to wheel up into line.——The facing and filing of the platoons will depend on which fide of the adjutant they are to be arranged, and which way the line is to face.

When a battalion open column entered and marching on a ftraight line is to form at a point, where its front flank is

to

to be placed, it will receive the word HALT when its leading division is at a wheeling diftance fhort of that point.

S. 125. *When the Battalion changes pofition by breaking into Open Column.—Marching up to the Point where its Rear is to reſt.—And entering the Line by the Wheeling of its*
Fig. 57. F. *Platoons.*

Befides the adjutant who marks the point of entry, two advanced points of March muſt be given.—The battalion then enters by Wheels and moves (as in S. 115.) and when its laſt divifion is at its point, it receives the word HALT, and pivots being corrected the whole are ready to wheel up into line.	Fig. 52. Wheel. Halt, Dreſs. March. HALT.

A battalion open column entering a new pofition where its rear flank is to be placed.—If the wheels are made to the pivot hand, it receives the word HALT when its rear divifion has juſt compleated its wheel into the new direction.—If the wheels are made to the reverſe hand it receives the word HALT when the laſt divifion but one has compleated its wheel into the new direction, and the laſt divifion itſelf files and places its pivot flank at the given point.——When a battalion open column entered and marching on a ſtraight line is to form at a point where its rear flank is to be placed, it will receive the word HALT when the pivot of its rear divifion arrives at that point.

P   By

By these operations of entering a new line at the rear, or at the front point, will the distant changes of a considerable line generally be made; each battalion breaking from the old line, and entering the new one in separate column; the whole of which movement may be made in quick time; the battalions within themselves are at all times collected, there can hardly be any impediments from ground (where it is possible for troops to move at all) that can prevent the transit of the battalion column from the one point to the other: the line is taken up just, by placing the pivot flanks upon it, and the distances are most correct, being taken up in all cases from the front of the column.—Should the presence and nearness of an enemy make it too precarious, thus to change position in detached columns; the ECHELLON March must then take place.

Fig. 57. H.

S. 126. *When the Battalion changes Position by breaking into Open Column.—Marching up in Column, and entering the new Position at the Point where a Central Division is to rest, and form in Line.*

It will often happen that the head of the battalion column must by wheeling enter the alignement at a point not far distant from where that head is to be placed in line; On its arrival there, the rear platoons cannot then have entered, but are stopped in the old direction by the cessation of movement in the front, it therefore becomes necessary immediately to bring those platoons into the alignement, that the battalion may justly form, and this is done by filing.

The

The leading platoon of the battalion having *wheeled* into the alignement followed by the others, when it arrives at the point where it is to form into line, the word HALT is given and the column stops.—The leading platoon and such others as may have already wheeled into the alignement being now at their proper points remain so, and the word FACE is then immediately given when all the platoons who are still in the old direction face to the flank which conducts to their place in the new line.

Wheel.
Halt, Dress.
March.
HALT.

Fig. 53.

— FACE.

At the word Q. MARCH, the serjeants mark their points in the line and the platoons move and halt with their pivot flanks on it ready to wheel up into line.

Q. MARCH.
Halt, Front.
Dress.

This movement includes both the operations of the battalion as entering a line where its rear is to rest, and where its front is to rest.

———

S. 127. *When the Battalion changes Position by breaking into Open Column.—Marching up in Column to the Point where its Head Division remains placed in the new position, and which its Rear Divisions enter on, by the Echellon March.*

Fig. 54. 78.

The column will advance to the spot where its leading division is to be placed: It will there receive the word HALT.——The leading division will if necessary be wheeled accurately

rately into the new line: Each of the other divisions will wheel back on its reverse flank such number of paces as is necessary to place it perpendicular to its point in the new line; the whole will MARCH and successively form up to the leading division by the echellon movement. (S. 158. 159.)

If the column halts perpendicular (A) to the new line, its divisions will wheel back 1-8th of the circle, or a half wheel.—If the column halts oblique (B) to the new line, the divisions will proportionally wheel, so as to be placed perpendicular to their future lines of march.

In this manner the divisions of the column arrive in full front, one after the other in the new line.

S. 128. *When the Battalion changes position by breaking into Open Column.—Marching in Column to the point in the new position where its Head is to rest, and to which its Rear Divisions form by successively passing*
Fig. 57. D, *each other and wheeling up.*

Fig. 55.

Wheel up.

Halt, Dress.

March.

Halt, Dress.

The column having arrived in the direction of, or in any direction oblique or perpendicular behind the line, and at the point where its head is to rest, but which its rear is to pass, its leading division will wheel into the line, and halt; each other division continuing its March will move on square behind the first formed division, at which point its leading officer will if necessary shift to its inward flank, and

and each as it comes oppofite to its ground will fucceffively wheel, march up and drefs in line with thofe already in it.

If the column is marching in the direction of the line, it will of courfe have its pivot flank on it, but as in this formation the wheel is made to the reverfe hand, therefore before it begins, the battalion muft fhift the breadth of the column to bring the reverfe flanks on the line and be directed by them, the leading officers at the fame time fhifting.

In this manner the battalion does not ftand in open column on the new line, but fucceffively wheels up by divifions and forms in full front on the given objects.—It may be ufed when the direction of its march is nearly in the prolongation of the new line, and when a battalion arriving on the flank of a line already formed, has to lengthen out that line.

S. 129. *When the Battalion changes pofition by breaking into Open Column.—Marching up perpendicular to the new Line, and to the point where its Head is to reft, and forming in the new pofition by the Eventail or Fan Move-*
Fig. 56. 57. E. *ment.*

When the leading divifion (b.) is at leaft the length of the battalion column behind its point in the new line, it fhortens its ftep one half as foon as the others receive orders to OBLIQUE from the column; this they do 'till oppofite their refpective

} TO THE—OBLIQUE.

Forward. ⎫ respective places, when each moves *forward* successively to
Half Step. ⎬ the leading platoon and to each other, take up the half step,
HALT. ⎭ enter the line in front, and the whole HALT.

This movement is performed on the March, and must be begun at a distance behind the line proportioned to the body which is to oblique and form.—It may be applied to one battalion, but hardly to a more considerable body, which would find great difficulty in the execution.—It gives a gradual encrease of front during a progressive movement.—With justness it can be made on a front division only, not on a central or rear one: In proportion as the leading platoon shortens its step will the one behind it, and successively each other, come up into line with it.—As soon as the colours of the battalion come up they become the leading point.

Although it is an operation of more difficulty, yet if the leading division continues the ordinary, and the obliquing ones take the quick step till they successively are up with it, a battalion column which is placed behind the flank of a line, may in this manner during the march, and when near to an enemy gradually lengthen out that line.

CHANGES OF POSITION OF THE OPEN COLUMN, MADE ON A FIXED POINT BY THE FILING OF COMPANIES.

Fig. 63. The changes of position of a column are the same as those of a line, after that line has broken into column.

S. 130.

S. 130. *When a Battalion in Open Column, changes position on a front fixed Company—by throwing forward or backward the Pivot Flanks of the rest of the Column.*

That company is placed with its pivot flank in, and perpendicular to the new direction, and points before it and behind it are given as directed for the battalion, the others FACE, MARCH and cover it in the new line. Fig. 47.

S. 131. *When a Battalion in Open Column changes position on its rear fixed Company—by throwing forward or backward the Pivot Flanks of the rest of the Column.*

Each company countermarches; the given company is placed.—The change then becomes the same as on the front company.—Each company again countermarches, and the column is in a situation to move on as before. Fig. 47.

S. 132. *When a Battalion in Open Column changes its position on any Central fixed Company.*

That company (a) is placed with its pivot flank in, and perpendicular to the new direction, and points (c. b.) before and behind it are given, where the pivots of its adjoining companies are to be placed: Fig. 48.
all

all such as were in front of it, countermarch and face it.—The whole then FACE to, and FILE from which ever (but the same nominal) flank is required in order to cover before and behind the placed company, and to arrive in the new direction.—The companies that face the placed one again COUNTERMARCH, and the column is in a situation to move on.

Should it be intended to form the line immediately after making the change of position.—In that case the company which faces to, and is next the placed one would take care to *Halt* in the new position, with a double distance from the placed one, and the line would be immediately formed by the WHEEL up of companies, without making the second countermarch.—It is always to be remembered, that whenever two platoons *face* each other in the same column, with intention to form in line they must have double distance, as they both *wheel inwards*, and meet on the line of formation.—But when the column after changing position is to be countermarched in part and proceed in the new direction, in that case no double distance is taken, and the necessary caution is given accordingly.

―――――

S. 133. *When a Battalion in Open Column changes to a distant position in its Front.*

Fig. 51 52 53. The column will march forward to some given point in that line, and then enter it, according to one of the prescribed modes, at which its head, central, or rear division is to stand.

―――――

S. 134.

S. 134. *When a Battalion in Open Column changes to a distant position in its Rear.*

Each division of the column will countermarch, and it will then proceed, as having the position in its front. Fig. 51. 52. 53

S. 135. *When a Battalion in Open Column changes to a distant position to either Flank.*

The companies will FILE from the old, into the new direction; or if the position is distant, the head of the column will march towards it, and enter it as a position in front. Fig. 49.

S. 136. *When the Battalion Column with the Right in Front is to form to the Right Flank.*

If the battalion is required suddenly to be formed on the ground on which it then stands.—The right pivots will quickly be covered, and the divisions will wheel to the right into line: in this situation the divisions of the battalion will be inverted.

If no inversion is to take place, the formation will be a successive one, by the head division wheeling to the right, and the others marching on past it, and successively wheeling up. (S. 128.)

Fig 60. 61.   When the head of a column advances and enters a new direction, by wheeling to its pivot hand, or by filing its divisions from its reverse hand, the formation made on that line by wheeling up the divisions, will front towards the rear of the column.—When the head of the column advances and enters a new direction by wheeling to its reverse hand, or by filing its divisions from its pivot hand, the formation made on that line by wheeling up will front the same way as the head of the column did when advancing to the line.

Fig. 78.   The open column forms in line on its front, rear, or central division by the Echellon march as in §. 158.

The open column closes to close column on any named division and forms in line by the deployments of the close column.

CLOSE COLUMN.

Application of the close column.   1. The battalion close column is formed from the column of march, or from line.—From the column of march it is generally formed for the purpose of assembly, or deploying into line.—From line it is formed in order quickly and in force to pass a defilé, or bridge: to make an attack in certain confined situations, where circumstances make it eligible: to oppose, in ground where its flanks are not protected, a threatened charge

charge of cavalry: to facilitate movements to the front, flanks, or rear, from which afterwards any other distances may be taken, or the line may be formed in the most expeditious manner.

2. The close column will generally be composed of companies for the purposes of movement: But when it is halted, and is to deploy into line, it will then stand two companies in front, and five in depth.

3. The same general circumstances apply to the close column as to the open column.—When the close column is formed: Rear ranks are one foot asunder, divisions are one pace asunder: Officers and serjeants are on the pivot flanks of their companies: Colours and supernumerary officers and serjeants are on the flanks not the pivot ones: Music, drummers, pioneers are ordered into the rear of the column: Artillery is either in the front, or on the reverse flank of the column when in march.

4. The commanding officer alone gives orders to the close column for its MARCH, HALT, and commencement of formation.

5. The battalion close column may be formed from line: in front, or rear of either of the flank companies: or in front, and rear of any central company.

Formation to front or rear.

6. *If the Column is to stand faced as the line is*, the battalion will face INWARDS, or to the directing company, each other company will disengage its head, march, and place itself as ordered before or behind that company.

7. *If the Column is to stand faced to the rear of the line,* then the directing company will countermarch on its own ground, the battalion will face OUTWARDS, or from the directing company; each other company will disengage its head, and move in file towards its place in the close column, by this means accomplishing a countermarch of the whole, and the column standing fronted to its former rear.

FORMATION OF CLOSE COLUMN FROM LINE.

Fig. 64.   *S.* 137. *Before, or behind either of the Flank Companies.*

CAUTION.

TO THE ——— FACE.

A CAUTION will be given mentioning the company, and whether the formation is in front or rear of it.—The battalion will then be FACED to that company, and the heads of the other companies will disengage to which ever hand, naturally conducts them towards their place in the close column: The officers and their covering serjeants post themselves at the head of their files ready to lead; the officer of the named company, shifts if necessary to that flank which is to become the pivot one of the column, and his serjeant also places himself 6 or 8 paces before or behind him (according to circumstances) to mark the perpendicular of the front of the column.

Q. MARCH.

The whole will MARCH QUICK to the front or rear of the company ordered to be formed on, and each leader will proceed

ceed in the same manner as in forming an open column from line, (except that the serjeants do not run out) stopping in his own person at his pivot point, and giving his words *Halt, front—Dress*, to his company, when it has arrived upon the proper ground on which it is to stand in close column.

Halt, Front. Dress.

During the formation of all close columns, as soon as the battalion is put in motion, the commanding officer will immediately place himself in front of the column, before the officer of the named company, and from thence judging the perpendicular of the column will attend to the officers covering each other in that direction as they come up, whether such covering is taken from the front or from the rear, which will depend on the formation of the column.

S. 138. *On a Central Company.*   Fig. 64.

A CAUTION of formation is given.—The named company will stand fast, and the battalion will face INWARDS; the heads of companies will disengage according as they are to be in front, or rear, the officer of the named company will place himself on its future pivot flank; and at the word MARCH, the rest of the formation will proceed as before directed, part of the battalion arranging itself before, and part behind the given company, and the officers covering on the proper pivot flank.

CAUTION.

INWARDS, FACE.

Q. MARCH.

Halt, front. Dress.

---

1. In

1. In the same manner, in which close columns are here formed from line on any given division by facing and disengaging; may columns at half or quarter distance also generally be formed: observing that in such cases the covering serjeants run out to mark their respective flank points, as in the formation in open column.

Fig. 65. 66.

OUTWARDS FACE.

2. In forming close column facing to the rear, the same operations take place, as to the front, with this difference.—That the CAUTION expresses what is to be done; that the named division *countermarches*; that the other divisions of the battalion FACE OUTWARDS from it, and lead from their farthest flanks, in order to establish the countermarch of the whole.

3. The close column is formed from column of march, by *halting* the head division, and ORDERING the others to close up, and *Halt* successively.——Or, by the head division continuing its March, and the rear ones being ordered to MARCH QUICK into close column, and successively to resume the ordinary march.

———

The close column marches to its flank: to deploy; to correct intervals; to gain an enemies flank; or for some other particular purpose: But a considerable movement to front or rear cannot be made without loosening its divisions and ranks.

S. 139.

§. 139. *When the Column marches to a Flank.*

A CAUTION will express to which flank it is to march: if to that which is not the pivot; the leading officers and serjeants of each will move quickly by the rear of their divisions to that flank; and the supernumerary officers and serjeants and colours who were on that flank, will exchange to the other.

COLUMN WILL MARCH TO THE ———

The whole will then FACE, and be put in MARCH, the officer that leads the front division taking care to march in the exact alignement, and all the others in preserving their proper situations, Dress and move by him.——When the column HALTS, FRONTS, the pivot officers and serjeants, &c. &c. are ordered to shift to their proper places (if not already there,) by the rear of their respective divisions.

RIGHT (OR LEFT) FACE.

Q. MARCH.

HALT, FRONT.

---

§. 140. *When the Column marches to the Front.*

The whole step off at the word MARCH, OR QUICK MARCH. If it is meant to loosen the ranks of the column, a CAUTION so to do will be given; on which all the divisions except the leading one will *step short*, and each successively from its leader will receive a word *step out* when his ranks are one pace asunder. If a general word HALT is given, the whole column

MARCH.

LOOSEN RANKS.

Step out.

HALT.

Halt.

{ column halts as it is then placed; but if a partial and low word *Halt* is given to the leading divifion only, the others ftill move on, and *Halt* fucceffively in clofe column by word from their leaders.

---

§. 141. *When the Column halted is to take a new Direction.*

CAUTION.

{ A CAUTION will be given that it is to change direction either to the right, or left; on which the officers and ferjeants if not already there, fhift to the flanks that are to lead.—The front divifion of the column is placed in the new direction, and an advanced point is given to determine the future line of pivots.

—— FACE.

Q. MARCH.

Halt, front, Drefs.

{ The other divifions will FACE as ordered, and MARCH quick.—Each divifion feparately when it arrives at its point which the ferjeants may give, will *Halt, front,* and cover in column.—Officers and ferjeants will again fhift to their pivot flanks if neceffary, and the covering of pivots be perfected.

---

§. 142. *When the Column marching, changes Direction.*

If gradual and inconfiderable changes of direction are to be made during the march of the column, the head will ftep fhort, and will on a moveable pivot gradually effect fuch change,

change, while all the other divisions by advancing a SHOULDER and inclining up to the flank which is the wheeling one, will successively conform to each other, and to the leading division, so that the whole at the word FORWARD may move on as before.

S. 143. *When the Column is to make Front to its Rear by Countermarching.*   Fig. 71.

If the divisions are at a sufficient distance, they will each separately countermarch as directed for the open column.— If the column is quite close, the whole FACE from the pivot flank; the even or every other division (reckoning from the head) will MARCH on till their rear has quit the column 3, or 4 paces, they then are ordered to COUNTERMARCH towards the column, and at the same word the odd divisions which have hitherto stood still, countermarch also each on its own ground: the even divisions march on 'till they are again in column in their proper places, and *Halt, front.*

THE COLUMN WILL COUNTERMARCH.
——— FACE.
EVEN DIVISIONS Q. MARCH.
HALT.
THE WHOLE COUNTERMARCH.
Q. MARCH.

Halt front.
Dress.

DEPLOYMENT OF THE CLOSE COLUMN INTO LINE.

1. The battalion close column forms in line, on its front, on its rear, or on any central division by the DEPLOYMENT, or flank march, and by which it successively uncovers and extends its several divisions.

2. Before

2. Before the close column deploys, its head division whether it is halted or in movement, must be on the line into which it is to extend.—That line is therefore the prolongation of the head division, and such points in it, to one or both flanks as are necessary for the formation of the battalion are immediately taken.

Attentions in the deploy-ment.

3. The flank March must be made, firm, marked, at the deploy step, parallel to the general line and without opening out, the most particular precision is therefore required.—Each division when opposite to its ground, will be most advantageously FRONTED, or at least corrected by a mounted officer of its own battalion, in case that its leader should not be critical in his commands, or that he should not be heard, or that his files are too open; and thus may the defects of a preceding division be remedied, by the judicious stop of the one following it: The division is then brought up into line by its respective leader.—— The justness of formation depends altogether on officers judging their distances, and timing their commands.——The officer who leads his division up into line must take great care that it does not overshoot its ground; his dressing is always from the last come up division, towards the other flank, and the eyes of all are turned to that division.

4. As the head of the close column is always brought up to the line on which it is to extend; therefore when the formation is made on the rear, or on a central division, such division when uncovered, must move up to the identical ground which the front has quitted.—The method formerly practised of throwing back such divisions as are before that of formation is improper, and will not apply where several battalions, or columns, are to form in the same line.

5. In

5. In the paſſage of the obſtacle, parts of the battalion are required to form in cloſe column, and again deploy into line, although the diviſion formed upon continues to be moveable.

6. Before any column deploys, the diviſions are well cloſed up and ſquare, and muſic, drummers, &c. are in its rear, or on the flank not the leading one, that the movement may not be embarraſſed.

When the Battalion close Column of Companies (the Right in Front) Deploys into Line.

S. 144. *On the Front Diviſion.*

Fig. 67.

The column being halted with its front diviſion in the alignement, and all the others in their true ſituations parallel, and well cloſed up to it, a point of forming upon and dreſſing is taken, in the prolongation of that diviſion (and corrected from it,) juſt beyond where the left of the battalion is to extend D.——A Caution is given that the line will form on the front diviſion. } FORM LINE ON THE FRONT DIVISION.

At the word to the Left face, the front diviſion ſtands faſt, its officer ſhifting to the right, and all the others face. —At the word March they ſtep off quick with heads dreſſed, } LEFT, FACE. Q. MARCH.

moving

moving parallel (not oblique) to the line of formation; the files alfo are clofe and compact, without opening out.

Halt, Front. Drefs.
March.

The officer of the fecond or leading divifion having ftepped out to the right at the above word March, allows his divifion led by his ferjeant to go on a fpace equal to its front, and then gives his word *Halt, front—Drefs,* his ferjeant ftill remaining on the left of the divifion.—He then being on the right of his divifion immediately gives his word *March,* and the divifion proceeds at the ordinary ftep towards its place in the alignement.—The officer having in the mean time ftepped nimbly forward, places himfelf before the left flank of the preceding divifion and is thus ready to give the word *Halt, Drefs,* at the inftant his inward flank man joins that divifion: He then expeditioufly corrects his men (who have dreffed upon the formed part of the line) on the diftant given point, and refumes his proper poft in line.

Halt, Drefs.

Halt, Front. Drefs.
March.
Halt, Drefs.

In this manner every other divifion proceeds, each being fucceffively (by its officer who himfelf ftops on the left flank of the divifion which precedes him) *Fronted, Marched up, Halted,* and *Dreffed* in line: The officers of thefe divifions as each approaches within 5 or 6 paces of its ground, then ftepping up to the flank of his preceding formed divifion, that he may the more accurately *Halt, Drefs* his own; and the flank ferjeant of each remaining at his point in the line, till the fucceeding officer having fo dreffed his divifion comes to replace him; he then covers his own officer.

S. 145.

S. 145. *On the Rear Division.*   Fig. 68.

The column being placed as before directed and a point of forming (D) taken to the right in the prolongation of the head division, and juft beyond where the right of the battalion is to come.

A CAUTION is given that the line will form on the rear division; on which the officers commanding divifions, and their ferjeants, immediately pafs behind their feveral divifions, and poft themfelves on the right of each; an under officer is fent from the rear divifion to place himfelf correctly clofe to and before the left flank file of the front divifion; and the leader of the front divifion is fhewn the diftant point (D) in the alignement on which he is to march, taking his intermediate points if neceffary.   FORM LINE ON THE REAR DIVISION.

The word to the RIGHT FACE is then given, on which all the divifions except the rear one, face to the right.—At the word MARCH, the faced divifions ftep off quick, the heads of files are dreffed to the left, the front one moves in the alignement, and the others parallel and clofe on its right.   RIGHT FACE. Q. MARCH.

As foon as the rear divifion is uncovered, it receives the word *March*; on this the divifion proceeds, and when within a few paces of its ground, its officer fteps nimbly up to the detached under officer who marks its left in the new pofition, he there in due time gives his words *Halt, Drefs,*   *March.*

Halt, Drefs.

and

{ and quickly corrects his division on the distant point of formation, this done he replaces his serjeant on the right of his division.

Halt, Front. Dress.

March.

Halt, Dress.

In the mean time the commander of the division which immediately preceded the rear one, having at the first word MARCH, stepped nimbly round to the rear of his division, without impeding its movement, and having allowed it to move on led by his serjeant, gives his words *Halt, Front—Dress,* when his division has marched a distance equal to its front, and thereby uncovered the one behind it, which immediately moves forward; he then places himself on its left, and his serjeant remains on its right.—As soon as his own front is clear he gives his word *March,* on which his division proceeds, he himself when proper, advances to the right of the preceding division then on the line, and from thence gives his words *Halt, Dress,* when his own left file joins such right; he corrects his division on the right, and then replaces his own serjeant.

Halt, Front. Dress.

All the other divisions succeffively proceed in the same manner, until the right one (which has been marching critically in the alignement, and on no account getting before it) receives when it arrives on its just ground the words, *Halt, Front—Dress.*

Fig. 69.

S. 146. *On a Central Division.*

Forming points (D. D.) must be given to both flanks in the

the prolongation of the head division.——At the CAUTION of forming on a central division, the leading officers will shift accordingly.—The divisions in front of the named one face to one flank; those in rear of it to the other, according to the hand which leads to their ground —The named division when uncovered moves up into line to its marked flank: Those that were in front of it proceed as in forming on a rear division: Those that are in rear of it proceed as in forming on a front division.

> THE LINE WILL FORM ON ——— DIVISION.
>
> OUTWARDS, FACE.
> Q. MARCH.

S. 147. *When the close Column of Companies forms Column of two Companies, or Grand Divisions.*

Fig 70. 72.

On the CAUTION, that the alternate companies from the front will form grand divisions, all supernumeraries, &c. but not the colours, go to the rear of the column if not already there.—At the word FACE, the alternate companies face (always to the pivot flank) and their officers then take one step sideways, so as to be clear of their rank.—At the word MARCH the officers stand fast, the serjeant of each conducts the division, and the officer of each when it has cleared the standing division, gives the words *Halt, Front—Dress*—*March* and *Halt, Dress* when he arrives at the one he is to join, his serjeant being on the flank of his division in the same manner as in deploying into line. The colours remain

> ALTERNATE COMPANIES WILL FORM COLUMN OF GRAND DIVISIONS.
>
> ——— FACE.
>
> MARCH.
>
> *Halt, Front.*
> *Dress.*
> *March.*
> *Halt, Dress.*

with

{ with their proper divifion in the column, and that divifion muft of courfe outflank on the hand not the pivot one.

<small>CLOSE DISTANCE TO THE FRONT.

MARCH.

Halt, Drefs.</small>

{ The officers and ferjeants now fhift their places, and take poft (whether the column has its right or left in front) fo that the right company of each divifion has its officer and his ferjeant on its right; and the left company has its officer on its left, and his ferjeant on its right, or in the center of the divifion.—A CAUTION is given to clofe diftance to the front,—The divifions move at the word MARCH, by the pivot flanks, and each pivot officer gives his words *Halt, Drefs*, when his divifion has clofed.—The clofe column is then ready to deploy or to march.

WHEN THE CLOSE COLUMN OF TWO COMPA-
NIES IN FRONT, IS TO DEPLOY.

S, 148. *On the Front Divifion.*

<small>THE LINE WILL FORM ON THE FRONT DIVISION.

——— FACE.

Q. MARCH.

HALT FRONT.</small>

{ The CAUTION of deployment is given, the line is prolonged, and attendant circumftances prepared.—The divifions that are to move, receive the word FACE, (always in this cafe to the pivot flank).—They move in file at the word MARCH.—A mounted officer gives, fucceffively and in due time, to each divifion the word HALT, FRONT.—The inward officer of each divifion when it has halted and fronted

gives

gives his words *Dress—March—Halt, Dress* and the outward officer assists him by remaining on the flank of the division in the line, in the same manner that the serjeant does for the company.———The left officer then replaces his serjeant on the right of his proper company.

Dress.
March.
Halt, Dress.

Fig. 67.

In this manner division after division comes up into line, and the supernumeraries, &c. also gradually take their places in the rear.

---

§. 149. *On the Rear Division.*

The CAUTION of deployment is given, the line is prolonged, and an under officer sent from the rear division to the pivot flank of the front one.———The divisions that are to move receive the word FACE (which in this case is always from the pivot flank).———They move in file at the word MARCH.

CAUTION.

—— FACE.

Q. MARCH.

The division that is immediately before the rear one, as soon as it has uncovered the rear one, receives from the mounted officer the word HALT, FRONT, and *Dress* from its inward or pivot officer; and at that instant the rear one is ordered to *March* forward by its pivot flank, and to *Halt*, and *Dress* in the line.—The division which preceded the rear one, and is now halted and fronted, when it is itself uncovered, in consequence of the movement of those before it, is also ordered to *March* forward and to *Halt*, and *Dress* in the line.

HALT, FRONT.
Dress.

March.
Halt, Dress.

March.
Halt, Dress.

S

HALT, FRONT. *Dress.* *March.* *Halt, Dress.*	In this manner each division as it uncovers the one behind it, successively HALTS, FRONTS by command from the mounted officer, and when it is uncovered is brought up into line by its own inward officer, aided by the outward officer.—This done the left officer replaces his serjeant, who has preserved his post in the front rank.

S. 150. *On a Central Division.*

CAUTION. OUTWARDS, FACE. Q. MARCH. HALT, FRONT.	The double operation of forming on a front and rear division is required.—The CAUTION of deployment is given.—The divisions FACE outwards—MARCH—and there must be an officer to HALT, FRONT, those of each wing.—The individual divisions proceed as already directed.

The column must always be well closed up, before it deploys.—When it deploys on a front division, it faces to the pivot flank which then becomes the leading one.—When it deploys on the rear division it faces from the pivot flank which then becomes the following one.

The close column when it forms on a front, or rear division may either be halted, or in motion to its flank.—From this situation of the flank march it is, that every battalion is required to begin the deploy when forming in line with others, and must therefore be much practised by the battalion when single.—Viz.—After the column has been placed in the alignment, it is FACED according as it is intended to form

on

on the front, or rear divifion, and is then put in MARCH, its head divifion following the alignement: At any inftant the divifion to be formed upon is ordered to HALT, FRONT, and the others without ftopping proceed and deploy upon it; if it is the front one it is already in the line; If it is the rear one the point which it comes up to, remains marked for it. When the formation is on a central divifion, it muft always begin from the halt of the clofe column.

The fingle battalion fhould alfo in exercife deploy on the front divifion when in march; as it is the method by which the line is reformed after paffing an obftacle—and of lengthening out the flank of a line that may be in movement.

OBLIQUE DEPLOYMENTS.

Although the quickeft, moft exact, and general method of deployment, requires that the battalion before deploying fhould ftand perpendicular to the line on which it is to form; yet it may fometimes happen that the immediate deployment of a column may be demanded, on a line oblique, to the one on which it then ftands, and that circumftances do not permit of the previous operation of placing it perpendicular to that line.

S. 151. *If the Deployment is to be made on an Oblique Line advanced.*

The front divifion is wheeled up into the new direction

Fig 73. B. on its REVERSE flank, and the line is prolonged to D.—The column is FACED to the hand it deploys to.—The leaders of divisions then turn their bodies so as each to take a direction parallel to the given one.——The whole are put in MARCH, and the rear of the divisions gradually get into the square direction of their heads which proceed and form as usual. In this movement the heads of the divisions will be a little retired behind each other: The rear leaders will take great care not to close on each other, nor to the hand which conducts them: much precision is also required in justly timing the HALT FRONT of each division, which by that time ought to be moving perfectly parallel to the line of formation.

S. 152. *If the Deployment is to be made on an Oblique Line retired.*

Fig. 73. A. The front division is wheeled up on its PIVOT flank, into the new direction, and the line is prolonged to D.—The same operation, though more difficult takes place as when the line is advanced, and the rear divisions must take particular care to ease from and yield to the march of the front.—The head division being advanced a few paces before it takes the oblique direction will give a facility to the heads of the rear files in gradually gaining it.

Such deployment can hardly be required on any other than the front division of the close column; particular attention is necessary to give

every

every aid as to the points of forming, and to the heads of divisions moving as soon as possible in the true direction parallel to and behind the line.—Should a column be ordered to form on the rear or on a central division; altho' the principles would be the same as on the front, and as in the other rear or central deployments; yet the execution would be very difficult, and demand great circumspection in the commanders of battalions.

S. 153. *When the Close Column Halted, is to form in Line in the Prolongation of its Flank, and on either the Front, Rear, or a Central Division.*

The CAUTION of Formation is given.—The named division stands fast, the others MARCH forward in close column in the given line: Their pivot officers successively take wheeling distance from each other beginning at the named one, and successively give their word *Halt* as each has acquired it: When the whole is in open column the line is formed by a wheel up to the flank.——In this manner distances are begun to be taken from the rear; but when the named division is a front, or central one, the others that are behind it must FACE ABOUT, MARCH forward, take their distances and *front* successively.   Fig. 74. A.C.

The column may also be opened from any named division, by the leading one only marching off, and each other successively following, as wheeling distance is acquired from the one preceding: When the whole have opened, the general word HALT is given, or the column is allowed to proceed.   Fig. 74. B.

ECHELLON

ECHELLON

CHANGES OF POSITION OF THE BATTALION BY THE MOVEMENTS OF THE ECHELLON COLUMN OF COMPANIES.

<small>Utility of the Echellon of march, in changes of position.</small>

1. The Echellon position and movements, are not only necessary and applicable, to the immediate attacks and retreats of great bodies, but also to the previous oblique or direct changes of situation, which a battalion or a more considerable corps already formed in line, may be obliged to make to the front, or rear, or on a particular fixed division of the line.

<small>How formed.</small>

2. The oblique changes are produced by the wheel less than the quarter circle of divisions from line, which places them in the Echellon situation.—The direct changes are produced by the perpendicular and successive march of divisions from line, to front, or rear.

<small>Fig. 75.

How applied.</small>

3. The march in line, or in the direct Echellon B, produces new parallel positions to front or rear.—The march in Echellon C, when formed by the wheels of the divisions from line, produces new oblique positions, to front, or rear, according to the degree of wheel given to the Echellon.—The march in open column A, produces new prolonged positions, to either flank.

<small>Echellon formed by wheels of companies.</small>

4. The Echellon of march, necessary in making changes of situation, will be composed of companies or sub-divisions, and generally formed from line by the wheel of each on its own flank, to the hand to which

it

it is to move.—Such wheel will feldom exceed the eighth of the circle, but can never amount to the quarter circle, otherwife the body would ftand in open column.

5. The Echellon of march may be confidered as a column of a particular kind, as well as the open column; and is eafily converted into fuch.

Echellon column.

6. All the divifions of an open column A, march upon one and the fame perpendicular, and are therefore eafily conducted.—All the divifions of an Echellon B, C move on different perpendiculars, each on its own but all of them parallel to the directing one, and removed from each other a fpace equal to what the divifions cut within each other.— In open column the perpendicular diftance from divifion to divifion, is equal to the front of the following one. In Echellon the fmaller the wheel is, the fmaller is the perpendicular diftance from divifion to divifion, till it vanifhes into nothing: but in all fituations of the wheeled Echellon, the oblique diftance from flank to flank, is equal to the front of the preceding divifion.—In open column the proper pivot flank is the directing one, and the wheels are made on it, into column backward and into line forward. In Echellon the reverfe flank (or that which firft joins its preceding divifion, when the line is to be formed forward,) is the directing one and the wheels are made on it, into Echellon forward, and into line backward.—In open column each divifion preferves a diftance from flank to flank, equal to its own front. In Echellon each preferves a diftance from flank to flank equal to the front of its preceding divifion.—An Echellon may at any time be converted into the open column, by wheeling up its divifions till they ftand perpendicular to the line which paffes through all its directing flanks. An open column

Differences and agreement of the open column, and Echellon.

Fig. 75.

lumn may in the fame manner be converted into the Echellon column, by wheeling back its divifions, each a named number of paces, and on either flank according to circumftances.

Method of forming Echellon by wheels from line.

7. The wheel from line into open column is eafily afcertained, by the perpendicular halt of each divifion on that line; but the parallelifm of the wheels into Echellon which is a circumftance that is effential, and decides the juftnefs of the movement, is more difficult to be determined; for being confined to no certain portion of the circle, fuch cannot well be announced or executed as a direction, and therefore *a given number of paces to be wheeled by bodies of equal ftrength*, and which ferve as fo many parallel bafes of formation, may be the beft general order that can be given.

8. If the companies of a battalion or more confiderable body were all of equal ftrength, and fhould the outward man of each take the fame number of paces on the circumference of the circle which he defcribes, they will after the wheel ftand parallel among themfelves: but if thofe companies are unequal, they will then not be parallel to each other, and confequently not in a proper relative fituation.——Tho' fuch equality may exift in a fingle battalion, it will feldom or never exift in a line of battalions, and a different calculation and direction for each battalion correfponding to their ftrengths, appears neceffarily to be required, whenever they are in concert to change pofition.——This difficulty may be obviated by adopting a *practical rule* as well for the battalion, as for the line, on all occafions of wheeling by companies into Echellon in order to change pofition, and of whatever ftrength the companies may be, viz. *That each covering ferjeant as the cafe requires,*

having

having previously placed himself, before or behind a given file (the 8ᵗʰ.) from the standing flank shall take the named number of wheeling paces, and thereby become a direction for the company to wheel up to, and halt: as in S. 154. 158.——As eight paces of the eighth file compleats the quarter circle or WHEEL, so four paces gives the HALF WHEEL, and two paces the QUARTER WHEEL, all which are wheels often made from open column, or from line, to change to a position perpendicular, or more or less oblique to the one quitted: and these degrees, with the helps given by advancing or keeping back a shoulder as is necessary, during the movement, will perhaps suffice to arrive and form in any new direction with precision.

9. The flank directing files of Echellons, whether they are formed by the perpendicular march of divisions successively from line to the front, or by the wheels of divisions from line to the flank; will at first, and should always afterwards be found in a diagonal line with respect to the front of divisions: In the first case A. the distance from flank to flank, depends on the interval which the divisions are ordered to march off at: In the second case B. such distance is always the same, and equal to the front of the division which has wheeled forward, and which by wheeling back would exactly fill it up.——Whenever therefore the directing flanks of an Echellon are all in the same line, and each distant from its preceding one, a space equal to the front of the preceding division, such Echellon is in a situation by wheeling back, to form in line to the flank, as in S. 156. or to take a position forward as in S. 162.

General situation of the directing files of Echellons.

Fig. 76.

10. In the Echellon march, such division or divisions as may meet

with

Paffing obstacles. with obftacles, will file round them without deranging the adjoining divifions who preferve the neceffary vacant fpaces, and diftances, till the broken divifions can again take their places.

Changes to the rear in Echellon. 11. When a change of pofition or march to the rear is to be made in Echellon—The battalion or line will in general FACE about, WHEEL into Echellon and then proceed.——Or—It may be ordered firft to WHEEL back into Echellon, then FACE about, and proceed as above.

CHANGES OF POSITION OF THE BATTALION FROM LINE, BY THE ECHELLON MARCH OF COMPANIES.

1. When the outward flank man of the company formed three deep, is ordered to wheel up three paces, or, if formed two deep, to wheel up two paces, fuch wheel is fufficient to difengage its rear rank from the front rank of the following one.—In fuch fituation a certain fmall degree of inclination may be gained to a flank in proportion to the front of the company which has fo wheeled, and the adherence of the feveral companies clofe behind each others flank, fhould facilitate the operation: but when a greater degree of inclination to the flank is required, then a more confiderable wheel up by companies is made, that each may thereby be placed in the perpendicular direction, which it is to purfue.

2. It has been obferved that the degree of wheel into Echellon, is always lefs than the quarter circle, and that the 8^n. file from the ftanding flank, is always the one to which the named number of wheeling paces

paces (33 inches each) is applied, in order to enfure the parallelifm of the companies however unequal they may be, and whether they wheel backward or forward.——Alfo that the degree of wheel made from line into Echellon, is always fuch as is required to conduct the divifions in a perpendicular direction to their future points; and this required degree muft be determined by trial, or by the eye of the commander, before he announces his order to HALF WHEEL—QUARTER WHEEL— or WHEEL any named number of paces, as 2. 3. 4. 5. 6. 7.

S. 154. *When a Battalion from Line—Wheels forward by Companies, to either Flank into Echellon and Halts.*

1. At the general CAUTION, that the companies will wheel forward fo many paces to the right or left, fo as to place them perpendicular to their future lines of march; the officer if not already there, moves to the named flank of his company, and the covering ferjeant of each at the fame time runs out, places himfelf before the 8ᵗʰ. file from the named flank, immediately takes the faid number of wheeling paces, on the circumference of the circle of which his flank man is the center, and then ftands faft with his body turned in the line of that flank man, who alfo faces into the line of his ferjeant.—The whole ferjeants ought thus to be in a line, but if any fmall correction is neceffary, it will immediately by the commanding officer be made from the leading flank.

COMPANIES, WHEEL FORWARD — PACES TO THE ——

Fig. 79. A.
Fig. 86. A.

t 2 At

Q. MARCH.

Halt, Drefs.

{ At the word MARCH, each company wheels up, till its 8ᵗʰ. file arrives clofe behind the ferjeant, at which time the officer who is on the ftanding flank gives his word *Halt, drefs*, eyes are turned towards him, and the dreffing being compleated the ferjeant places himfelf on the outward wheeling flank.

In this fituation the flanks wheeled to remain in an exact line, and alfo the wheeling flanks if the divifions are of equal ftrength: but in proportion to the degree of wheel which has been made, will the perpendicular raifed from the ftanding flank of each divifion cut within the divifion preceding it, till by the compleat wheel of the quarter circle all fuch perpendiculars coincide, and beyond that, new Echellon fituations begin to the rear.

2. When the movement is to be to the rear inftead of the front; in that cafe the battalion will in general FACE to the right about and WHEEL forward into Echellon in the before manner, proceeding as if the line was to its proper front.

3. Or, The battalion may occafionally be firft ordered to WHEEL BACK into Echellon (as in S. 158.) and then to FACE ABOUT, and MARCH to the rear: they thus do not ftand for any time unneceffarily faced to the rear, previous to the operation of marching, which is a circumftance to be avoided as much as poffible.

S. 155.

S. 155. *When a Battalion having from Line wheeled into Echellon—Marches forward and Halts; ready to form in such Direction as shall be required.*

The companies standing thus parallel to each other, and their leaders being on the pivot or flanks wheeled to.—At the word MARCH, the whole move on at the ordinary step, each flank on its own perpendicular; each officer is now attentive to preserve the distance he marched off at from his preceding pivot, and also his oblique covering in the line of pivots, which remain always parallel to the original line; this requires the greatest care, being an operation more difficult than moving in open column, where all the pivots cover each other in the same line.——These circumstances observed the Echellon may at any instant be ordered to HALT, and will then be in a situation ready to form up, parallel or oblique to the line it quitted. If parallel, by each division wheeling back to the flank of the one immediately behind it. If oblique, by the divisions moving up into the direction which the leading one then has, or is to be placed in, as is hereafter directed.

MARCH.

Fig. 79.
Fig. 86. A.

HALT.

The Echellon can at no time march in any other direction, than in the one to which it stands perpendicular, except that an oblique march of the whole divisions should be required from it.—During the march, the same great regulating circumstances that direct the open column,

direct

direct the Echellon, viz. the preservation of distance from the preceding leading flank, and the diagonal lining or covering of all those flanks, at the same time that the perpendiculars of march are preserved by each division.——Could the march in Echellon be always executed with the greatest accuracy, each flank leader covering a certain file of his preceding division at a certain distance, would ensure exactness: but this alone is not to be trusted to, and is rather to be considered as an aid, than as an invariable rule; for the unsteady or open march of one or more divisions, if productive of a waving or shifting of the following ones, would in a sensible manner influence the whole.——If the leaders of the two head divisions do preserve an equal and steady pace under the direction of the commanding officer, who keeps close on the flank of the first one, and gives such directions to the second, as are necessary for preserving the parallelism of the march; those two will serve as a base line, on which all the others should cover.——In this as in every other case, the perfect perpendicular march of the first leader in consequence of his body being truly placed, and his attention solely given to this object, is what will much determine the precision, and justness of the whole.

———————

S. 156. *When the Battalion having Wheeled from Line into Echellon, has Marched, and Halted—and is to form back, parallel to the Line it quitted.*

WHEEL BACK INTO LINE. A CAUTION is given that the companies wheel back into line.—On which the pivot men face into the line, and the officers

officers take one step forward.—At the word MARCH, each company wheels back to the new pivot, and on receiving from its officer the word *Halt, Dress,* eyes are turned towards him.—The line being completely formed, officers and serjeants (if not already there) move to their respective places in line; except in the occasional case of wheeling into line, in the middle of a change of position.—For officers do not then shift from their leading flanks (unless ordered) but remain there ready to fire, and to wheel again into Echellon to resume the march, when the supposed sudden attack of cavalry is repulsed.

Q. MARCH.

Halt, Dress.

Fig. 79. B.
Fig. 80. B.

S. 157. *When the Battalion having Wheeled from Line, into Echellon has marched and halted—and is to form up, oblique to the Line it quitted.*

Fig. 77.

Various circumstances attend the execution according to the degree of wheel which must be given to the leading company in order to place it in the required oblique position, and as the number of paces which have been already wheeled from line into Echellon, determine the nature of the Echellon, they are an essential part of the following arrangements.

1. *If the formation is made forward;* and the leading company is wheeled up the same number of paces, that it before wheeled from line into Echellon: then the others without altering their situation move on, and successively dress up with it.—

Fig. 79. C.
77. B.

In

In this manner does one or more battalions make their changes of position on a flank or central company of the line.

Fig. 77. C.
2. *If its wheel up exceeds* that number of paces; the others wheel up one half of that excess, move on, and successively dress up with it.

3. *If its wheel up is less* than that number of paces; the others wheel back one half of what they originally wheeled forward, after deducting one half of what the leading division has now wheeled forward; they then move on, and dress up with it.

4. *If the formation is to be on the prolongation* of the front division as it stands; the others wheel back one half of what they originally wheeled forward, then move on, and dress up with it.

5. *If the leading division has to wheel back* into the new position; the others wheel back (in addition to the one half of what they originally wheeled forward) half of what the leading division has now wheeled, move on, and dress up with it.

All these specified wheelings are in order to make the divisions stand perpendicular to the lines by which they must march to their points of formation, which lines change in consequence of the position given by the leading division.

S. 158.

S. 158. *When from Open Column, the Companies Wheel backward into Echellon, in order to form in Line on the Head Company.* Fig. 73.

The head company either remains square to the column, or is wheeled forward on either flank into the intended direction of the line, and on the position given it, will depend the relative one which is taken by the other companies, and which the commander will determine to himself, by his eye, or by immediate trial.

1. On the CAUTION, that the companies except the head one will wheel back on the right or left so many paces (and which wheel is always backwards, and always on the reverse flank of the column, as being that which afterwards first comes into line,) the officer moves to that flank, and the serjeant of each places himself with his back to the 8th. file of the rear rank, immediately takes his named paces, and halts fronts with his body turned in the line of the flank man on whom he wheeled.——At the word MARCH, the company wheels back till the 8th. file of the rear rank touches the breast of the serjeant, (who gives a low caution to halt) it is then halted and dressed by the officer from the standing flank, the serjeant places himself on the outward flank, and the whole are now in a situation to march forward, and form in line on the head company, as in S. 159.

COMPANIES, WHEEL BACKWARD, ―― PACES, ON THE ――.

Fig. 73. A. C.
Fig. 54. A. B.

MARCH.

Halt, Dress.

2. *If the line was to be formed on the rear company of the co-lumn;*

lumn; that company would remain placed; the others would FACE ABOUT—wheel BACK on the pivot flanks of the column, as being those which afterwards first come into line —MARCH,—and then *Halt, front* successively in the line of the rear company.

3. *If the line was to be formed on the rear company, but facing to the rear*: The whole column would first countermarch, each company by files, and then proceed as in forming on a front company.

4. *If the line was to be formed on a central company of the column:* That company would stand fast, or be wheeled on its own center into a new required direction.—Those in front of it would be ordered to FACE about.—The whole except the central company would WHEEL back the named number of paces; those in its front on the proper pivot flanks of the column, and those in its rear on the reverse flanks, such being the flanks that first arrive in line.—The whole would then MARCH into line with the central company, as in S. 161.—If the column was a retiring one, and the line was to front to the rear the divisions must each countermarch before the formation begun, and the head would be thrown back and the rear forward.

Fig. 78. B.

S. 159. *When the Battalion changes Position to the Front, on a fixed Flank Company, by throwing forward the rest of the Battalion.*

When

When the commander has determined the new line to be taken, by placing a perfon, a, in it, 20 or 30 paces beyond the fixed flank; he orders the ferjeant from before the 8th file of the flank company, to wheel up into that line, thereby to afcertain the number of paces required.—He then directs that company to be wheeled and halted in the new pofition, and the adjutant to prolong the line as far as the moving flank of the battalion will extend.

Fig. 77. A.

Fig 86. C.

The CAUTION is then given to the other companies, to wheel towards it, half the number of paces, that the flank one has done, for thereby will each ftand perpendicular to the line, which is drawn from its flank in the old line to its relative flank point in the new one, and it is along fuch line that each will move.—The battalion wheels into Echellon as in S. 154.

COMPANIES, WHEEL FORWARD ——— PACES, TO THE ——

Q. MARCH.
Halt, Drefs.

The officer being on the inner, and the ferjeant on the outer flank of each company, the whole except the fixed company will move on at the word MARCH as directed in S. 155.

MARCH.

When the officer conducting the fecond company approaches within 7 or 8 paces (and not fooner) of where his leading flank is to join the firft company already placed, he gives a word—*Shoulder* (the outward one) *forward*, on which the man next to himfelf preferving the fame ftep gradually turns his fhoulder, fo as to arrive on the new line fquare in his own perfon; and the reft of his divifion (who till this inftant have marched in their original perpendicular direction)

—*Shoulder forward.*

Halt, dress up. — tion) conforming to him and proportionally lengthening their step, arrive in full parallel front on the line, so as to have a very small movement to make at the word *Halt, dress up*, which is given by the officer when his leading flank touches the flank of his preceding company: he himself having nimbly stept forward when at 3 or 4 paces distance, and being then before that flank, instantly halts his men, and corrects them on the distant given point, their eyes being turned towards him, and the formed division.

—Shoulder forward.

Halt, dress up.

In this manner company will come up after company (or division of whatever kind after division) each following one, observing to give the word *—Shoulder forward*, when the preceding one gets the word *Halt, dress up*, and each officer stepping up to before the flank of his preceding formed company when he is within 3 or 4 paces of it, that he may the more quickly and accurately give his word *Halt, dress up* to his own men, which they are to do preserving the cadenced step of the division, and not suddenly springing backward or forward.—The serjeants will remain in the line till they are relieved by the officers whose places they occupy.

The exact formation in this oblique line depends totally on the companies having wheeled (only) one half of the angle which the new position makes with the old one, for should they at first wheel the whole of that angle, they would be then marching parallel to that line and arrive in it doubled behind each other; where as by having the other half of the wheel to compleat when they come near to the new position,

tion, each moves in a perpendicular direction, and disengages the ground required by the succeeding one to form upon.

S. 160. *When the Battalion changes Position to the Rear on a fixed flank Company, by throwing backward the rest of the Battalion.* Fig. 77. B.

The new position is given, and the flank company wheeled into it in the manner already directed, but backwards instead of forward.

The rest of the battalion FACES to the right about, the companies then wheel forward the given number of paces towards the standing flank—or—as is already mentioned, they may if so ordered wheel BACKWARD into Echellon, and then FACE about.

} RIGHT ABOUT FACE.
COMPANIES FORWARD
WHEEL —— PACES
TO THE ——
Q. MARCH.
Halt, Dress.

The companies MARCH with their rear ranks in front, and form in line in the same manner as when changing position forward; except that the officer of each, having timeously given his word *Shoulder forward,* when his preceding one *Halts, fronts,* and then having disengaged himself from his division, will as soon as his leading flank man of his front rank touches the preceding formed flank, give his word *Halt Front, Dress back,* on which his company fronts, and without hurry dresses back on him and the formed part of the line; he correcting them upon the more distant given point.

MARCH.

—*Shoulder forward.*

Halt front.
Dress back.

Very

Very great activity is required from the officer in dressing up, or dressing back, otherwise the point of appui will not be ready for the next officer who arrives and is to perform the same operation, and this will particularly happen where the change of direction is inconsiderable.—In the successive dressing of divisions in this manner officers are always to line them, so as not to obscure the distant point, but to leave it open and distinct, so that the direction of the line may run at the distance of one file from the given object of dressing.

S. 161. *When the Battalion changes Position on a Central Company, by advancing one Wing, and retiring the other.*

Fig. 80.

1. The central company is wheeled into the new position as already directed, and backwards or forwards according to the wing it belongs to.—Two points D, D, are quickly taken in the line, about where the flanks of the battalion are to extend. and in the line of the central company.

—WING, RIGHT ABOUT FACE.
COMPANIES WHEEL—
FACES INWARDS.
Q. MARCH.—*Halt, dress.*

{ The retiring wing FACES about—both wings WHEEL their companies inwards and forwards, half as many paces as the central company wheeled.

MARCH.

{ The whole MARCH forward into line with the central company, the advancing wing dressing up, and the retiring wing fronting and dressing back, as already directed.

2. *During*

2. *During the march of divisions to the front, into a new direction*, if they should be obliged to form in line in order to repulse a sudden attack of cavalry; the whole will HALT; the inward or directing flank of each will stand fast, and the outward one instantly WHEELS back to its succeeding one; when the enemy is repulsed, the march is resumed, by each company WHEELING up its outward flank to its former position, and then proceeding in the movement—during this operation, the officers remain on their Echellon flanks, from thence halt, dress them when they wheel back into line, fire them if necessary, and from thence also wheel them again into Echellon.

> HALT.
> WHEEL BACK INTO LINE.
> Q. MARCH.
> *Halt, Dress.*
>
> WHEEL INTO ECHELLON.
> Q. MARCH.
> *Halt, Dress.*
> MARCH.

3. *During a march to the rear*, if this operation is necessary—The whole HALT, FRONT; each company instantly WHEELS up its outward flank to the pivot preceding, and the line is thus formed, officers remaining on their Echellon flanks.—When the movement is to be resumed, the whole FACE to the rear, each company again WHEELS forward its outward flank the required number of paces as at first, and the MARCH is continued.

> HALT, FRONT.
> WHEEL UP INTO LINE.
> Q. MARCH.
> *Halt, Dress.*
>
> RIGHT ABOUT FACE.
> WHEEL INTO ECHELLON.
> Q. MARCH.
> *Halt, Dress.*
> MARCH.

4. *If the change of position is a central one:* Then both the above operations may take place at the same time. The general situation, if the whole is at any period of the movement halted and formed will be: such central part as has arrived at the new line will be formed in it; but the flank parts which have not entered, and which join each of its extremities will be formed in lines parallel to each other and to the position which they quitted.—When one **flank** only is required to form

and halt, the other will continue to purfue its proper formation in the new line.

5. The fquarenefs of each Echellon and individual, and the perfect equality of ftep during the movement are what alone can produce the decided exactnefs required in thefe operations.

S. 162. *When the whole Battalion being moveable, changes Pofition to front or rear, on a diftant point, which is in the Interfection of the old and new Line.*

Fig. 77. 79.

1. *If the change is made to the front;* every company is wheeled up to the leading hand, half the number of paces and no more, that would be required to place it parallel to the new pofition; the whole move on in their perpendicular direction till the flank of the leading company arrives in the new line: it then immediately wheels up as many paces as it before wheeled, and halts, dreffed in the new direction.—The other companies march on, and as they fucceffively arrive near the new line, they advance their outward fhoulders, and halt, drefs in it.

2. *If the change is made to the rear;* the whole face about, and break into Echellon.—Each again fronts and dreffes back when it has arrived in the new line.

This is the movement performed by each of the battalions of a confiderable line, except one flank or one central one, in moft changes of pofition made on a point within the line.—For there can be but one

battalion

battalion of a line which forms on a fixed division; all the others are evidently moveable forward or backward; each in proportion to its distance from the general center, and from the point where its leading flank is to rest in the new line.

S. 163. *When from Line, the Companies of a Battalion, march off in Echellon, successively and directly to the front, and again form in Line, either to the front, or to the flank.*

1. *As long as the intention is to form to that front;* they may be retired at any named distance whatever behind each other, and when the leading division *Halts*, the others may move on, and dress in line with it. Fig. 81. A.

2. *But when the intention is to form in line to the flank;* the whole will be ordered to HALT, or the divisions successively to take any named distance and *Halt.*—The directing flank of the leading company will be considered as the first point in the intended oblique line, and the particular direction meant to be given it, will be established by the placing of another point (a,) beyond and before it.—A serjeant from each company will run out, and post himself as a pivot, lining on the first given points, and on each other, each also taking a distance from the one before him equal to the front of the division which precedes him. Fig. 81. B, C.

The rear companies are then, by the oblique march to their directing hands, or by facing and filing should situation require it, marched to their respective serjeants, and then *Halt front*, square to their former front.

front.—The line is formed by the WHEEL back of each company on thofe eftablifhed flanks—Or—When the companies are thus placed, the whole may be put in MARCH to the front, and preferving their relative fituations HALT and WHEEL back into line, at fome more advanced point.

ECHELLON CHANGES BY SUBDIVISIONS OR SECTIONS.

In the Echellon movements by companies in order to gain ground to a flank, and afterwards to make a parallel, or an advanced oblique formation: If the wheel up of each is confiderable, it becomes the more difficult to preferve the true diftances during the march, and thereby to refume the parallel line when fo ordered, by the wheel back of companies.——In many cafes therefore, fuch changes if not limited to fixed points may be made by the fubdivifions or fections wheeling up three paces only, fo that each can afterwards move forward independant and juftly, by remaining clofe behind each other; and this may be done either when the battalion is halted, or when it is in motion, without the intervention of advanced ferjeants, or difplacing of officers, but merely by the regular wheel up of the divifion with its outward man, who takes the three ordered paces.

S. 164.   *If the Battalion is halted.*

CAUTION. { 1. The CAUTION is given that the fubdivifions or fections will wheel 3 paces to right or left.

At

At the word MARCH, the outward man of each fubdivi- | Q. MARCH.
fion or fection, whatever its ftrength may be, wheels up 3 |
paces, and each company officer gives the word *Halt, Drefs,* | *Halt, Drefs.*
to the ftanding hand: in this fituation the divifions will |
ftand parallel or nearly fo, and the front rank of each will |
be immediately behind the line of the rear rank of its pre- | Fig. 8. A.
ceding one: the 3 file of colours and center ferjeants will | Fig. 88. A.
wheel up as a feparate divifion parallel to the others. |

2. After the wheel of fubdivifions, the company officer will be on the pivot flank of his firft one, and his ferjeant on that of his fecond. After the wheel of fections to the right, the company officer and his ferjeant will be on the right of the 2 leading ones, and an officer or ferjeant from the rear on the right of the others.——After the wheel of fections to the left, the leading ones will have an officer or ferjeant from the rear on their left, and the company officer and his ferjeant will be on the left of the two laft ones.

3. A fubdivifion of 9 files that wheels up in this manner 3 paces, will ftand at an angle of about 30 degrees with its former front, and if it is only formed 2 deep and wheels up 2 paces, it will ftand at an angle of 20 degrees.——A fection of 5 files that wheels up 3 paces will ftand at an angle of about 35 degrees.——According to the ftrength therefore of the divifion that thus wheels, will be the degree of obliquity taken from the former pofition.

At the general word MARCH, the whole move on in their |
then perpendicular direction as fpecified in the Echellon |
movements, taking care that the ftep is equal, and that each | Fig. 83.

keeps

MARCH.	keeps up to its preceding division, but by no means throws forward its advanced flank, which would necessarily derange the others; and should one division commit this fault, the succeeding one ought not to be influenced by it, but still maintain its equal step, and thereby avoid a shake or hurry in the rear by which distances and direction would be lost.
HALT.	The whole halt.

WHEEL BACK INTO LINE. Q. MARCH. *Halt, Dress.* Fig. 88.	4. *If the battalion is to resume its former front* B, C.—It instantly receives the CAUTION to wheel back into line.—At the word MARCH, each division wheels back 3 paces, thereby joining the next standing pivot, and immediately receives the word *Halt, Dress,* from the leader of each company wherever he may be (always on the right to which if necessary he will have shifted) to whom the whole of his company then turn their eyes, and are by him corrected on the standing pivot.

HALT. FORM BATTALION FORWARD. MARCH.	5. After the HALT: *If the battalion is to form forward,* in the direction B, D. of its leading division, or that that division is previously wheeled up into a more advanced one.—The company leaders will shift if necessary, each to the flank of his leading division if in sections.—A CAUTION is given to form battalion, and at the word MARCH, the whole except the head division move on, and each pivot leader of the front rank by a small and gradual turn forward

ward of his inward shoulder (if necessary) conducts his division at an equal pace towards the point of each in the new line, and when within 5 or 6 paces of it, by the bringing forward the outward shoulders the division arrives in it on a parallel front, where each receives from the company leader (who is then on the moving flank of his company) a word *Halt, Dress*, and to which each successively conforms as he repeats it for them; in this manner the subdivisions or sections will successively arrive in line, observing the circumstances of movement already prescribed. When the line is formed company leaders if necessary shift to the right of their companies.

Halt, Dress.

The whole of these movements depends on the accuracy of step, and the gradual and insensible turn of the shoulders of the pivot leaders, to which the divisions conform and by which they are conducted on the march, and into the new line.

S. 165. *If the Battalion is in March in Line.*

1. The intention being to gain ground to the flank by the Echellon march of subdivisions or sections and without making a previous halt.—On the word subdivisions or sections 3 paces to the right or left wheel; the pivot men of the front rank of each division, turning in a small degree to the pivot hand, mark the time for 3 paces, during which the named divisions wheel in ordinary time on those men, and the 3 file

Fig. 88.

SUBDIVISIONS ——
3 PACES, RIGHT
WHEEL.

FORWARD.	file of the colours and center ferjeants alfo wheel up as a divifion, parallel to the others.———At the 4th. pace and at the word FORWARD, the whole move on direct to the front which each divifion has acquired, the pofition of leaders being as already defcribed.
WHEEL BACK INTO LINE. FORWARD. HALT.	2. Where fufficient ground has been taken to the flank,—on the word WHEEL BACK INTO LINE, the pivot men mark the time for 3 paces, turning back in a fmall degree to their original front, and the fubdivifions or fections inftantly wheel backward into line, without altering the time and at the 4th. pace, the whole ftep on, having received the word FORWARD, till the battalion is ordered to halt.

HALT. FORM LINE FORWARD. HALT.	3. When fufficient ground has been taken to a flank, and —*that a forward formation of the line is to be made*, the head divifion halts in its then pofition, or is wheeled up 2, or 3, paces more and halted—The reft of the battalion receives a CAUTION to form on the head divifion, they continue their march, and conforming to the directions given in S. 164. by the gradual alteration of their fhoulders arrive fucceffively in line.

When the battalion is on two ranks only, two paces will be fubftituted inftead of three in all thofe movements—and no unneceffary time need be loft or paufes made, betwixt the execution of thefe feveral words of command.

---

4. In

4. In thefe cafes the original wheel up of divifions, being limited to the 2 or 3 paces which difengages them from each other, the inclination of their flank movement, or of their change of front, or pofition alters with the ftrength of fuch divifion.—If therefore a fmall degree is to be taken, they may fo wheel by companies, if a greater by fubdivifions, and if a greater ftill by fections, the clofe adherence of each to each enfuring (if well executed) the regularity of the battalion during the operation, which is made on the principles though without all the formality of the exact Echellon, and may be required and ufed in many fituations of movement and changes of pofition.

MARCH OF THE BATTALION IN LINE.

The MARCH of the battalion in LINE either to front or rear being the moft important and moft difficult of all movements, every exertion of the commanding officer, and every attention of officers, and men become peculiar neceffary to attain this end.—The great and indifpenfible requifites of this operation are; the direction of the march being perpendicular to the front of the battalion as then ftanding; the perfect fquarenefs of the fhoulders and body of each individual; the light touch of the files; the accurate equality of cadence and length of ftep, given by the advanced ferjeants, whom the battalion in every refpect, covers, follows, and complies with.——If thefe are not obferved, its direction will be loft; opening, clofing, floating, will take

<small>General attentions.</small>

take place; and diforder will arife in whatever line it makes a part of; at a time when the remedy is fo difficult, and perfect order fo effential.

Directing ferjeants.

Fig. 11.

It is evident therefore that every individual fhould be well prepared for this operation.—But more particularly to enfure its correctnefs, two or more directing SERJEANTS muft be trained to this peculiar object, on whofe exactnefs of cadence, ftep, fquarenefs of body, and precifion of movement dependance can be had.—The habitual poft of the two directing ferjeants in the battalion is to be, in the center of the battalion and betwixt the colours, one of them in the front rank, and one in the rear, that they thereby may be ready to move out, when the battalion is to march, one other alfo covers them in the fupernumerary rank.

At all times when the battalion is formed in line, and halted, the inftant attention of the front directing ferjeant is (after being affured that he himfelf is perfectly and fquarely placed in the rank) by cafting his eyes down the center of his body, from the junction of his two heels and by repeated trials, to take up, and prolong a line perpendicular to himfelf, and to the battalion; for this purpofe he is by no means to begin with looking out for a diftant object, but if fuch by chance does prefent itfelf in the prolongation of the line extending from his own perfon, he may remark it: he is therefore rather to obferve and take up any accidental fmall point on the ground, within 100 or 150 paces, intermediate ones cannot be wanting, nor the renewal of fuch as he afterwards fucceffively approaches to, in his march.——In this manner he is prepared under the future correction of the commanding officer to conduct the march.

S. 166.

S. 166. *When the Battalion halted, and correctly dressed is to advance in Line.*

The commanding officer having previously placed himself 10 or 12 paces behind the exact line of the directing serjeant, will if such file could be depended on, as standing truly perpendicular to the battalion, and great care must be taken to place it so, remark the line of its prolongation and thereby ascertain the direction in which it should march; but as such precision cannot be relied on, he will from his own eye, readiness, and having the square of the battalion before him, make such correction, and observe such object a little to the right or left, as may appear to him the true one; and in doing this, he will not at once look out for a distant object, but will hit on it, by prolonging the line from the person of the directing serjeant to the front: or he will order the covering serjeant to run out 20 paces, and will place him in the line in which he thinks the battalion ought to advance.—The directing serjeant then takes his direction along the line which passes from himself betwixt the heels of the advanced serjeant, and remarking his object preserves such line in advancing.

Fig. 82.

The commanding officer will give the CAUTION, the battalion will advance, on which the front directing serjeant moves out 6 accurate and exact paces in ordinary time and halts; the 2 other serjeants who were behind him, move up on each side of him, and an officer from the rear replaces in the front rank the leading serjeant. The center serjeant in moving out marches and halts on his own observed points, and

THE BATTALION WILL ADVANCE.

and the two other serjeants dress and square themselves exactly by him.——If the commanding officer is satisfied that the center serjeant has moved out in the true direction he will acquaint him so, if he thinks he has swerved to right or left he will direct him to bring up the shoulder on that side the smallest degree possible, in order thereby to change his direction and take new points on the ground, towards the opposite hand.

MARCH.

The line of direction being thus ascertained; at the word MARCH the whole battalion instantly step off, and without turning the head, eyes are glanced towards the colours in the front rank: the replacing officer betwixt the colours preserves during the movement his exact distance of 6 paces from the advanced serjeant, and is the guide of the battalion.——The center advanced serjeant is answerable for the direction, and the equal cadence and length of step; to these objects he alone attends, while the other two scrupulously conforming to his position, maintain their parallelism to the front of the battalion, and thereby present an object to which it ought to move square: they are to allow no other considerations to distract their attention, and will notice and conform to the direction of the commander only, and if any small alteration in their position is ordered, it must be gradually and coolly made.

Officers.

1. OFFICERS in the ranks can only be observant of their own personal exactness of march, they are then but individuals equally attentive

tive as their men; they are not to attempt to drefs their companies by looking along or calling to them, otherwife they will certainly err themfelves and derange the march: fuch care belongs to the officers in the rear, and well trained foldiers themfelves know the remedy that is required and will gradually apply it.

2. The weight of the COLOURS, and the embarraffment attending them in windy weather, rough ground, &c. make it impoffible at any time to depend on the officer carrying them for a true direction or an equal and cadenced ftep: but they muft always be carried uniformly and upright, thereby to facilitate the moving and dreffing of the line. Colours.

3. The MEN are on no account to turn their heads to the colours, but to preferve them and confequently their fhoulders fquare to the front, and to depend principally on the light touch of the elbow, together with an occafional glance of the eye, and the accuracy of ftep for their dreffing—If heads were permitted to be turned to the center, the inward fhoulder would be brought forward, the wings would remain behind, the files would open, and diforder would arife in endeavouring ftill to adhere to the center, and to counteract what would be occafioned by the fault of principle, and not of the foldier. Soldiers.

4. Inattention, or an inequality of STEP will produce a waving in the march of the battalion; but the communication of this may often be ftopped, by the exertions of the major and adjutant who feeing where and why it originates, will immediately apprize the companies in fault, and coolly caution the others that are well in their true line not to participate of the error.——A flank of the battalion may at firft fight appear to be behind, when the fault really arifes from a central Step.

tral divifion bulging out, and thereby preventing the flank from being feen.

5. Whatever ALTERATION is to be made in any part of the battalion muft be made gradually, and not hurried, that the confequent fhake it occafions may be as little felt as poffible; the mounted officers only can point out, and correct fuch faults.

Flanks.

6. The FLANKS on no account are to be kept back; much lefs are they to be advanced before the center; in either cafe the diftance of files muft be loft, and the battalion will not be covering its true ground: the convex or concave fhape of the battalion will fhow this to the commander, and the beginning of each inaccuracy is to be ftudioufly corrected by neceffary cautions.—The officer who is on each flank of the battalion, being unconfined by the ranks and not liable to be influenced by any floating that does arife; may, by preferving an accurate ftep, and having a general attention to the colours and to the proper line which the battalion fhould be in with refpect to the advanced directors, very much affift in preferving the flanks in their due pofition: When he obferves that a line drawn from himfelf through the center of the battalion paffes confiderably before the other flank, he may conclude himfelf too much retired; when fuch line paffes behind that flank, he may conclude himfelf too much advanced, he will therefore regulate himfelf accordingly.——When the battalion in march is convex, the wings muft gain the ftraight line of the centre, by bringing up the outward fhoulder; and it muft be ftrongly impreffed on the foldier, that in all fituations of movement, by advancing or keeping back the fhoulder as ordered, the moft defective dreffing will be gradually

dually and smoothly remedied, whereas sudden jerks and quick alterations break the line and produce disorder.

7. The REAR RANKS which were closed up before the march begun, must move at the lock step, and not be allowed to open during the march; the correct movement of the battalion depends much on their close order. *Rear ranks.*

8. Supported ARMS are allowed when halted or when in column as not interfering with its exactness; but in the march in line, arms are always to be carried SHOULDERED, as otherwise it is in vain to look for a just line, or true distances of files, and slovenliness, inaccuracy, and disorder, must take place at a time when the most perfect precision is required. *Arms carried.*

9. The COMMANDER must himself attend to the correct movement of the directing serjeant; if during the first 20 paces he perceives steadiness and no floating in the battalion, he may be assured that the line of march is justly taken; but the contrary will be the case if (the parallel front of the battalion being preserved) he sees the files on one flank opening and on the other crowding; he will instantly apply the remedy by ordering the directing serjeant—RIGHT SHOULDER FORWARD, if the opening is on the left of the battalion, or, LEFT SHOULDER FORWARD, if the opening is on the right: At this command the serjeant making an almost imperceptible change of his position (by bringing up one shoulder) and of his points, and the colours in the battalion when they have advanced 6 paces to his ground conforming to it, the whole will by degrees gain a new direction.—Every change of direction made in this manner must produce a kind of wheel of the *Change of direction on the center in march.* *Fig. 82. D. E.*

battalion

battalion on its center) one wing gradually giving back, and the other as gradually advancing, an attention which the commander must take care is observed.

HALT.
> The battalion marching in perfect order, when it arrives at its ground receives the word HALT; the step which is then taking is finished and the whole halt; eyes remain turned towards the center, the whole remain steadied, and the commanding officer places himself close to the rear rank, in order to see whether the battalion is sufficiently dressed, and in a direction perfectly parallel to the one it quitted.—No preparatory caution is to be used before halting, such caution supposes and encourages incorrectness and creates uncertainty: at the word halt, the whole halt firmly.

When the battalion is advancing in line for any considerable distance or moving up in parade, the music may be allowed at intervals to play for a few seconds only, and the drums in two divisions to roll, but it is the wind instruments only which play, the large drum, or any other instrument whatever which marks time by the stroke, is not to be permitted.—When the line is retiring, music are never to play.

S. 167. *When the Battalion is to Dress.*

It is evident that in the DRESSING of a single battalion after the halt, whatever correction is necessary must be made by advancing or retiring the

the flanks, and not by moving the center which having been the guide in the march, has juftly ftopped at the point where it has arrived.

1. When the commanding officer gives the word DRESS, the company officer on the left of the colours inftantly dreffes the 6 or 8 files to the right of the colour in a proper parallel direction, the two wings immediately conform to the center and afterwards receive the word EYES, FRONT.   DRESS.

EYES, FRONT.

2. Should the commander require a more exact dreffing than the above gives, he will order one colour to advance one ftep, and FACE to the left, alfo the fecond company officer on the left of the colour to advance one ftep, and FACE to the left; then the flank company officers to advance and to face to the center; then each other company officer inftantly to COVER thofe at their due diftances and face to the center; then the officers of the left wing to FACE about, fo as the whole ftand fronted to the left.——Then battalion RIGHT DRESS, on which the companies MARCH up to their refpective officers who are favourably pofted for halting and dreffing each his company; after which and without lofs of time the officers front into line.   Fig. 82. G.

BATTALION, RIGHT, DRESS.

MARCH.

Halt, Drefs.

3. It muft be obferved in this mode of dreffing, whether it is taken from the center or from a flank, that platoon officers who originally face to the left, take diftances equal to the front of their own platoons from the officer before them; but fuch as face to the right, muft take diftances from the officer before them equal to the front of the platoon which in line is on the right of them.——When circumftances allow the

dreffing

dreffing to begin from the left an advantage arifes, that the officers do all originally face to the left.

<small>Change of direction on the flank halted.</small>
4. A fmall change of direction may in this manner be given to the battalion when halted, either on a flank, or central company.—To the *Front* by advancing and placing the officers.—To the *Rear* by the covering ferjeants in the fame manner giving the ground, the men facing about, lining with the ferjeants; then fronting; and the officers replacing the ferjeants.—But a flank is never in fuch cafe fuppofed to move above 20 or 30 paces.

The battalion may alfo be occafionally dreffed in the following correct manner.—One of the colours is advanced fome paces.—An under officer on one flank of the battalion, is placed in a determined line.—An under officer on the other flank lines himfelf with the laft placed one, and the colour.—The two center grand divifions are moved up to the colour and dreffed to each flank.—The wing grand divifions then move up, and the grenadier and light company in fame manner. This dreffing may foon be made if done at the ordinary pace, without hurry, and that the chiefs of divifions aligne in the prolongation of the bafe.

S. 168. *When the Battalion is to retire.*

It is evident that it ought to be previoufly dreffed with the fame correctnefs, as when it was to advance, and the fame care in afcertaining the direction of its march muft be taken.—Therefore before the retreat is to begin an officer will have placed himfelf 30 paces in the rear,

rear, so as to stand perpendicular to the front directing serjeant, and of course he will be in the line, or nearly so, of the directing serjeants.

At the word, the BATTALION WILL RETIRE, the directing serjeants face about. The same center serjeant that directs to the front, directs also to the rear; he moves on in the line of the advanced officer, 6 paces beyond the rear rank and halts; and the other serjeants are on each side of him. } THE BATTALION WILL RETIRE.

At the word RIGHT ABOUT FACE, the whole face; and the supernumerary officer who replaces the directing serjeant, moves up into the leading rank; a mounted field officer passes through to the rear, and the directing serjeant in the interim prolongs his line, and takes his objects betwixt the feet of the posted officer. } RIGHT ABOUT FACE.

Immediately after facing about, the word MARCH is given, and the whole proceed in the same manner, and with the same attentions as in moving to the front; the directing serjeant conducting on his points, under the correction of the field officer who is 10 paces behind the battalion. } MARCH.

When the battalion is to front: it receives the word HALT, FRONT, and immediately halts and fronts, the serjeants, &c. resuming their proper stations; it is then dressed if necessary in the manner already prescribed. } HALT, FRONT.

1. In marching to the REAR, the battalion must cover its proper

Z

extent

Attentions in retiring.

extent of ground.—The rear rank men muſt avoid cloſing their files more than uſual, otherwiſe the front men who are in general larger, will be crowded in their rank.—Muſic, drums, ſupernumerary officers, &c. will take care to march with exactneſs, not to interrupt, but rather to aſſiſt the battalion.—The battalion is not to FACE about, till every thing is prepared for its inſtant MARCH, and its HALT, FRONT is one command: when retiring therefore it never unneceſſarily ſtands faced to the rear.

Wing platoons.

Fig. 83. A. B.

2. When the WING companies of a battalion are wheeled backward and faced outward in order to cover its flank.—Such companies if during the retreat they march in file, will take particular care to move in the ſame direction as the battalion and not impede its progreſs.—When the battalion fronts, thoſe companies will face outward, and always recollecting that their immediate buſineſs is to cover the flanks, they will regulate their poſition and movements by thoſe of the battalion.—When marching they move in file perpendicular to the line of the battalion: when fronted they make an angle with it of about 45°. according to the apparent circumſtances that threaten.

S. 169. *Changes of the Battalion when in Movement.*

The battalion when marching in front, muſt be much accuſtomed to ſtep out, to ſtep ſhort, to oblique to right or left, and to change direction by a ſmall and gradual turn of the ſhoulder: all theſe muſt be executed with the utmoſt preciſion, in perfect cadence, and upon decided

cided words of command, as they are operations wanted and essential to the perfect movements of a considerable line.

---

1. Obliquing a battalion in a parallel direction to gain a flank, or to preserve a given appui is a difficult, but necessary operation.—Obliquing a battalion when in line with others for a few paces in order to correct an interval, must be done without eyes being turned from the center. } RIGHT OBLIQUE. FORWARD.

2. Change of direction on the march begins with the leading serjeant, and is conformed to by the center and by the battalion, when they arrive at the point where the serjeant begun it; it must be made almost insensibly, and gradually in proportion to the extent of the body, that is thus to change direction, for without incurring disorder, the outward flank can only get into line by lengthening its step which requires time. } LEFT SHOULDER FORWARD. RIGHT SHOULDER FORWARD.

3. Obliquing the battalion by the wheeling up of sections or subdivisions is performed as in S. 164. and is used where a considerable space is to be gone over.

4. Change of front, and position by subdivisions or sections is performed as in S.165.

5. If the battalion halted or in movement is required to make a wheel on a flank with an uniform front, such wheel can seldom be wanted to exceed the 8th. or 6th. of the circle.

—On

RIGHT, WHEEL, FORWARD. HALT.
{ —On the word, to the right wheel; the right marks the time, the center takes a half step, and the left a full step; the intermediate parts of the battalion conforming accordingly, and at the word forward, or halt, the whole are directed by the center; this movement requires every aid that can be given by the mounted and supernumerary officers.

6. If the battalion in movement is required to make a small change of front on the center.——The center will mark the time, a very small turn of the shoulders will be gradually made, the wings will conform, one advancing the other giving back, till at the word forward the whole move on as before.

Wheeling up or throwing back the whole or part of the battalion.

Fig. 87.

7. A battalion halted may change its position forward to a certain degree, or throw back a flank; in a manner that gives great protection during the movement if made near an enemy.—At the word MARCH, the right company (or left) *wheels* into the new direction, and the rest of the line at the same time moves on in front, and by command OBLIQUES to join the left of the first company.—When the right flank of the second company has arrived there, it also *wheels* up into the new direction, and the rest of the line continues to oblique to join its left flank. In this manner the line preserves its uniform front, obliques, and gradually enters the new position as its leading company arrives in it, at the same time that it covers and protects the flanks of the formed companies.—By the same means also will a battalion throw back any number of its divisions in presence of an enemy: The angular company

company will give the direction, the reft FACE ABOUT. MARCH, OB- Fig. 89.
LIQUE; fucceffively *wheel* into it and *front*.—The outward company of
all which may be formed as a flank to the battalion, will march in file,
and cover the flank.——Before this movement commences officers muft
fhift to the inward flank of their companies, in order to drefs them on
the given diftant point after the wheel, in the fame manner as in the
Echellon movements, of the nature of which this partakes.

PASSAGE OF OBSTACLE WHEN THE BATTALION IS MARCHING IN LINE.

When the battalion is marching either to front or rear, the partial *Paffage of*
obftacles that prefent themfelves will be paffed, by the formation, *the obftacle in clofe co-*
march, and deployment of the clofe column.—Such parts as are not *lumn, either*
interrupted ftill move on in front; fuch parts as are interrupted, dou- *in advancing or retiring.*
ble by divifions as ordered, behind an adjoining flank or flanks, and
in this manner follow in clofe column in their natural order.—As the
ground opens they fucceffively deploy, and again perfect the line.——
The columns are always behind the line, and march clofed up.—The Fig. 84.
formed part of the battalion whether advancing or retiring continues
to move on at the ordinary pace, and in proportion as the obftacles
encreafe or diminifh, will the formed or column parts of the line encreafe or diminifh.

In general the columns formed will be of fub-divifions: the firft fub- *General*
divifion that is obliged to double will be directed to which hand by the *attentions.*
commander of the battalion, the others as they fucceffively double will

in confequence place themfelves behind it and behind each other, and the hand firft doubled to will be that which prefents the opening moft favourable to the fubfequent march, and formation, and which the commanding officer will always hold in view, and order accordingly. —The interrupted body will double to one or both flanks, according to circumftances, and the order it receives. Obftacles that impede a flank will occafion a fingle column to be formed from the flank towards the center. Obftacles that impede the center or a central part of a wing, will if confiderable, occafion two columns to be formed, from the center towards the flanks.——The columns will follow a flank of fuch part of the line as is not impeded; and either in doubling into column or extending into line, the rear divifions will conform to the movements of their then leading one.—No part lefs than the front of the column doubles or moves up, and when half or more of a battalion muft be thrown into one column, it will be ordered by companies.

S. 170. *When the Obftacle prefents a confiderable Front parallel to the Line.*

Fig. 84. C. D, &c.

CAUTION.

HALT.
FACE.

The divifions impeded muft all at once double behind fuch one or two other divifions, as clear them of the obftacle.——In this cafe a timely caution is given by the commanding officer to the part of the line that is to pafs the obftacle; the neceffary portion of the line when within a few paces of where it is impeded, is ordered to HALT, FACE either to one or both flanks, and the heads of the fub-divifions (except the leading one) difengage to the rear. The whole

whole MARCH quick, and each as it arrives square and close behind the preceding one *halts, fronts* and *marches* forward, taking up the *ordinary* step, when closed up. The leaders of the sub-divisions of the column, remain on the flank next the opening which they are to fill up.

Q. MARCH.
Halt, Front.
Q. March.
Ordinary.

S. 171. *When a Point of the Obstacle is presented to the Line, and that it continues to encrease.*

The doubling is then successive, beginning with that division which is first interrupted, and continuing as it becomes necessary till the column can advance in clear ground.— In this case the sub-division impeded will be ordered by the commanding officer to HALT, FACE, MARCH—*Halt, Front March* by its own leader, and follow the one adjoining to it, which makes the flank of that formed part of the line. When this last subdivision also becomes impeded, these two perform the above operation, and place themselves in column behind the next sub-division.—The three, the four, &c. successively repeat it as the narrowing of the ground requires (and upon the words given by the commanding officer, or by the officer of the then head division, should the commanding officer be otherwise employed) until the obstacle ceases to interrupt the march of a formed part of the line.

Fig. 84. G.
CAUTION.
HALT.
FACE.
Q. MARCH.
Halt, Front.
Q. March.
Ordinary.

S. 172.

S. 172. *When the Obstacle is passed, or diminishes, and that the Line encreases.*

HALT.
FACE.
Q. MARCH.
Halt, Front.
Q. March.
Ordinary.

{ If it is of such a nature as to permit of the compleat extension at once into line: the whole column performs it by the commands and deployments of the close column on the front division which then makes part of the line.

Fig. 84. T. H.

HALT.
FACE.
Q. MARCH.
Halt, Front.
Q. March.
Ordinary.

{ But when the obstacle diminishes by degrees only; then the divisions of the column must come up into line successively as the ground opens, and the remainder of the column must in diminishing shift towards the obstacle, in the same manner that it before shifted from it in encreasing. When the second sub-division of the column can therefore come up; its leader or the commanding or mounted officer gives the word for his own and the following sub-divisions HALT, FACE, MARCH, and when opposite to his ground HALT, FRONT, MARCH, and when he is up in line ORDINARY.—It depends on the opening of the ground whether more than one division of the column can come into line at the same flank movement.—This operation is repeated by the mounted officer, or the leader of what is then the second sub-division, as often as such sub-division sees that it is proper to move up into line, and is conformed to by the rear of the column till all its divisions have successively arrived in the line.

The commanding officer himself or a mounted officer must as much as possible order the doubling of the divisions, and their moving up into line; and particularly when any confiderable part of the battalion is obliged to double into one column.—But if there are feveral doublings in the battalion at the fame time, he can only direct the moft confiderable one, and the others muft be ordered by their feveral head officers.

Thefe movements are all made on parallel and perpendicular, not oblique lines, and the progrefs which the formed part of the battalion is conftantly making, fhows that no time muft be loft either in giving or executing the words of command, and that the divifions of the column muft be well clofed up, and its movements quick, firm, but in perfect order.—The divifions of the column form fucceffively into line, as the obftacle permits them, or again double fo as to conform to the fhape of the ground, which muft always be filled up.—The march of the uninterrupted part of the line muft be fteady and exact, and the openings made muft be carefully preferved from the center while it continues to direct, or from whatever point does fo while the center is impeded; the columns depend on the formed parts of the battalion to which they are attached, and are independent of each other.—When the center is interrupted, a named company officer of the line will be ordered to advance 6 paces to regulate the whole 'till the directing ferjeant of the center can again refume his true and original line, which he by advancing fingly from the column will endeavour to do as foon as poffible.

General attentions.

Whether the battalion is advancing or retiring the fame operations take place, and the columns in both cafes are behind the formed part of the line: in retiring the rear rank leads.

S. 173.

S. 173. *When the Batialion fires, during the Passage of an Obstacle.*

HALT.

{ If the battalion in *advancing* should be obliged to fire; it HALTS in the situation it is then in, executes such firings as are ordered, and again advances.

HALT, FRONT.

RIGHT ABOUT FACE. MARCH.

{ If the battalion in retiring is pressed by the enemy the part in line will HALT, FRONT, the part in column will move on 'till the last division arrives in line and will then HALT, FRONT. The firing that is ordered will be executed; and when it is again proper to retire, the whole will FACE about, the part in line will MARCH, and the columns will also be put in MARCH when the line arrives at their head.

S. 174. *When a Battalion is advancing it may also under certain Circumstances pass such Obstacles as present themselves, by File.*

In such case the interrupted division or divisions will be ordered to FACE either to one or both flanks, and closely to follow in file such parts of the battalion as are not broken: the filing will encrease as the obstacle encreases, but as it diminishes, file after file will successively and quickly move up to their proper place 'till the whole are again formed; and during this operation, the leading file will always remain

attached

attached to the flank of the part in line.—The same rules that direct the doubling in column, direct the doubling by files; when a subdivision files it will be from one flank only; when a company files it may be from both flanks; and if a larger front than 2 companies is interrupted, it then doubles into column.—Where the obstacles are of small extent, but frequently occurring, this mode is the readiest that can be applied in advancing: but in retiring it cannot be used, if the enemy are at hand to press upon the battalion; and therefore the passing by column is to be looked upon as the general method.

Fig. 84.
I, K, L, M.

In plate 10. fig. 84.—The position A. is a battalion and part of two others formed in line, they advance meeting with obstacles.—B, three sub-divisions of the left of the battalion have doubled.—C, one sub-division of the right has also doubled.—D. a central obstacle now occurring three sub-divisions of the right and one of the center, also two of the left, and three of the center have doubled.—E. one of the right has moved into line, and one more of its center has doubled, also one of its left has moved into line.—F. the whole divisions have moved into line, except three sub-divisions of the left which are in column.——G. the three sub-divisions of the left remain in column, and two on the right and three at the center have again doubled.—H. the whole having arrived on open ground have moved up into line, except one sub-division on the right.

If the battalion A. instead of advancing in front, is supposed to have faced to the right about and to be retreating; the positions of the divisions in column will be the same as above, they performing their movements, with their rear ranks in front.

The positions I, K, L, M, shew the passage of obstacles, by the impeded parts filing round them.—At N, the line is again compleated.— O, P, shew the passage of a wood, by the filing of companies.——At Q. all obstacles are passed; and the situation of part of the adjoining battalions also appears during this march.

S. 175. *When the Battalion moving in Line, passes a Wood, or other impediment, to front or rear, by the filing of Companies.*

PASS TO THE FRONT.

Right, turn.

Fig. 84. O. P.
Fig. 85. A.

1. *If to pass a wood or other embarrassed ground to the front;* when it is found necessary to break the battalion, the commander will order it to PASS from the right of companies to the front, on which each company officer orders his company *right turn,* wheels out his leading file, and passes on as fast as the difficulty of the ground will allow him, endeavouring to preserve a relative distance from the left as being the head of the column, or from the other flank if particularly so ordered.——Each officer on arriving at the farther edge of the wood will *halt* his company, and remain till the others are come up, and till the whole are ordered to march out, and form in battalion; which will generally be done by standing in open column the left in front, dressing pivot flanks, and wheeling up into line.—Or, if the companies form separately on the edge of the wood, they will march out and join in the battalion.

Fig. 85. B.

2. *If to pass to the rear.*—When the battalion retiring in line,

line, arrives at the point where it muſt break, it is ordered to, PASS, COMPANIES by FILES.—The leader of each gives his word *left turn*, and proceeds as above directed; the heads of files are regulated from the left; and after quitting the wood, at an ordered diſtance, they HALT FRONT into column, the right in front, and WHEEL to the left, up into line.—The line then again retreats if neceſſary.

PASS COMPANIES BY FILES.
Left turn.
Q. March.

HALT, FRONT.
&c.

3. *If a battalion in firſt line paſſes through a ſecond which advances and relieves it.*—The ſecond marches up to within 12 paces of the firſt and halts.—The battalion of the firſt then receives the word PASS COMPANIES BY FILES.—Each leader gives his word *Right face, Q. march*, and proceeds at a quick pace to the rear through the ſecond line, which, wherever the head of a diviſion preſents itſelf, throws back as many files as are neceſſary to give it paſſage, and again immediately moves up; the retiring files who are regulated by their left, at any ordered diſtance HALT FRONT into column the right in front, and WHEEL up to the left into line.

Fig 91.
PASS COMPANIES BY FILES.
Right, face.
Q. March.

HALT, FRONT.

4. *When the ſecond line does not advance to relieve the firſt.*—The battalion of the firſt line retires, and when it comes within 12 paces of the ſecond, it then receives the word to PASS COMPANIES BY FILES; each leader orders to the *left turn*, and proceeds as before directed; the column when halted and fronted, having its right in front.

PASS COMPANIES BY FILES.
Left turn.
Q. March.
&c.

Circumſtances may require, that the companies ſhould PASS from their proper left inſtead of the right, in which caſe

the

the leaders will shift and conduct such left, until the line is formed, when they will again resume their proper places.

5. *If a battalion in second line passes by files to the front, through a first line.*—It will advance within 12 paces of the first one.—On the command to PASS to the front by files; each company leader will give his word *right turn*, and move on at the head of his file in ordinary time, through the first line, which makes openings for it.—When the rear of the files has passed; the battalion will be ordered HALT FRONT in column the left in front—WHEEL into line—and may then advance.

There may be occasions, where instead of halting in column, and wheeling into line.—The battalion may be ordered to form by the rear files moving up to their front leaders; but the line thereby obtained will generally be a very inaccurate one, and not fit to advance without a halt, and a previous dressing.

6. *If a battalion in second line advances and passes in front, through a first line which it is to relieve.*—The first line will at the necessary instant wheel back by companies into open column, the advancing battalion will pass through it, such files as are interrupted following to the right, moving up as soon as they can, and the battalion thus reformed moving on to its object,——or,—if a battalion advancing in front meets with a line retiring, this last will throw itself into open column, and halt, till the advancing battalion has passed——or,—if a line is retiring in files, it will in same manner halt square when it meets the advancing line, allow it to pass and then proceed.

S. 176.

S. 176. *When the Battalion retires by alternate Companies in two Lines.*

The right companies stand fast, or, halt front if the battalion is already in motion.—The left retire in line a given number of paces and halt, front: on which the right companies retire in the same manner beyond the left, and halt, front.—In this way they proceed till the battalion is ordered to form.———One colour remains on the flank of its proper company in each line and directs its movement, for which purpose a serjeant will advance 6 paces before it, during the march. Distances are preserved from that colour.—The eyes of each line remain turned to their colour, and officers are on the inward flanks of their companies.—Each line has a commander.—The light infantry may be divided in the intervals of the first line, retire with it, and change to the other line, whenever it becomes the advanced one: in this situation they cover the retreat and may occasionally fire.

RIGHT COMPANIES HALT, FRONT.
LEFT COMPANIES HALT, FRONT.

RIGHT COMPANIES. { ABOUT FACE. MARCH. HALT, FRONT.

LEFT COMPANIES. { ABOUT FACE. MARCH. HALT, FRONT.

S. 177. *When the Battalion advances, or retires by half Battalions, and fires.*

1. If the battalion is in march and advancing.—The left wing HALTS when ordered, and the right one continues to move on 15 paces, at which instant the word MARCH being given to the left wing, the right at the same time is ordered to

LEFT WING { HALT. MARCH.

RIGHT WING. { HALT. READY. PRESENT. FIRE. MARCH. LEFT WING — HALT. READY. &c.	to HALT, to fire and load, and the left marches paſt them, till the right wing being loaded and ſhouldered receives the word MARCH, the other wing HALTS, fires, &c. and thus they alternately proceed.
RIGHT WING, HALT FRONT. LEFT WING, HALT, FRONT. RIGHT WING { READY. PRESENT. FIRE. ABOUT FACE. MARCH. LEFT WING, HALT, FRONT. LEFT WING. { READY. PRESENT. FIRE. &c.	2. If the battalion is in march, and retiring.—The right wing is ordered to HALT, FRONT, and when the left one has gained 15 paces, and receives the word HALT, FRONT, the right wing is inſtantly ordered to FIRE, to load, to FACE about, and MARCH 15 paces beyond the left, where it receives the word HALT, FRONT, on which the left wing gets that of FIRE, and in the ſame manner alternately proceeds, every due diſpatch being made in reloading.

There muſt be a commander for each half battalion.

One colour remains on the inward flank of each half battalion, to which the men continue to look, by which they move and before which a directing ſerjeant advances 6 paces.

The make ready, preſent, fire of the advanced wing is inſtantly to ſucceed the march of the other advancing wing, or, the halt front, of the retiring wing.

In the half battalion firing, advancing and retreating.—If formed 2 deep, both ranks will fire ſtanding. If formed three deep, the front, and center rank fire ſtanding, and the rear rank remains ſhouldered in reſerve.

———

S. 178.

S. 178. *When the Battalion forms a Square, or Oblong.*

1. The 4th, 5th, 6th battalion companies stand fast (in consequence of the explanatory caution that is given preparatory to forming the square) the rest of the battalion faces inwards, and disengages the heads of companies to the rear; the colours and their coverers fall back, the 4th company closing to the left to fill up their place.—They march quick.—The 7th, 8th, and light companies place themselves in open column behind the 6th.—The 3d, 2d, and 1st place themselves in open column behind the 4th, the grenadiers place themselves between the light company, and the 1st.——When these three last companies close up to the 8th and 2d, and face about (having each first countermarched, if it is thought necessary to have the front rank outermost) at the same time that the 7th, 8th, and the 3d, 2d, wheel outwards, the oblong stands compleat,—or,—the square may be a perfect one, if it is composed of the eight battalion companies only; the grenadier and light company, being in reserve in the rear, ready to be applied according to circumstances.

> CAUTION.
> FORM SQUARE.
> COMPANIES INWARDS FACE.
> Q. MARCH.
> *Halt, Front.*
> Fig. 92.

2. *The square or oblong may be formed, by the 4th, 5th, 6th, companies standing fast.*—The rest of the battalion wheels backward, each company the 8th of the circle on its inward flank.—They face about—They march to compleat the square as above; each wheeling when it comes to its ground, and then fronting; and in this manner will the proper front rank

> CAUTION.
> FORM SQUARE.
> COMPANIES BACKWARD WHEEL.
> Q. MARCH.
> *Halt, Dress.*

RIGHT ABOUT FACE. MARCH. *Halt, front.* *Dress.* Fig. 90.	rank of the rear face be outward.——The commanding officer, colours, and their coverers, drums, &c. &c. are within the square, as also the battalion guns, which are shifted to wherever they are most necessary.—The square is composed of the front, the right, the left, the rear faces: the front face is that on which the square originally forms.

THE SQUARE WILL MARCH, TO FRONT, REAR, RIGHT, OR LEFT. MARCH. Fig. 93. A. HALT. FRONT, SQUARE.	3. *When the square or oblong is to march by any one face.*—The side which is to lead is announced; the colours move up behind its center; the opposite side faces about; and the two flank sides wheel up by sub-divisions so as to stand each in open column.—The square marches; two sides in line and by their center; and two sides in open column, which cover, and dress to their inward flanks on which they wheeled up, carefully preserving their distances.—The square halts, and when ordered to front square, the sub-divisions in column immediately wheel back, and form their sides, and the side which faced about again faces outwards.

THE SQUARE WILL MARCH, BY THE RIGHT FRONT ANGLE. Fig 93. B.	4. *When the perfect square is to march by one of its angles in the direction of its diagonal.*——A CAUTION is given by which angle; and the two sides that form it stand fast, while the other two sides face about.—The whole then by sub-divisions wheel up one eighth of the circle, 2 sides to the right, and 2 sides to the left, and are thus parallel to each other, and perpendicular to the direction in which they are to move, the pivot flanks being in this manner placed on the

sides

sides of the square.——Each side being thus in Echellon, and the colours behind the leading angle, the whole are put in march, carefully preserving the distances they wheeled at, and from the flanks to which they wheeled.——After the HALT, and at the word FRONT SQUARE, the whole wheel back into square, and the two sides that require it face about outward.——When the oblong marches by one of its angles, its subdivisions perform the same operation of wheeling up, each the eighth of the circle; but its direction of march will not be in the diagonal of the oblong, but in that of a square, viz. of the line which equally bisects the right angle.

MARCH.

HALT.
FRONT, SQUARE.

The angular march of the square or oblong may be made in any other direction to right or left of the above one; but in such case the subdivisions of two opposite sides, will have to wheel up more than the eighth of the circle, and those of the other two sides proportionally less, in order to stand as before perpendicular to the new direction, the sum of these two wheels will always amount to that of a quarter circle, and their difference will vary as the new line departs more or less from the equal bisecting line; this will be known by first wheeling up the 2 angular divisions till they stand perpendicular to the new direction, and then ordering all the others to conform accordingly.—This movement is very difficult in the execution, and cannot be made with any degree of accuracy, unless the perpendicular situation of the divisions is correctly attained, and carefully preserved.

Fig 93. C.

5. *The*

5. *The square halted changes direction on any one of its sides;* by that side wheeling up, on one of its flank divisions; which is previously placed; its two flank sides at the same time make a similar gradual change to comply with the alteration; and the rear side marches in file to compleat the square.

---

6. *When the square in march halts and fronts, to repulse an expected attack of cavalry.*—The front rank kneels, and present their bayonets sloped; the two rear ranks fire standing; either companies by ranks succeessively; or companies (independant of each other) by subdivisions, one firing when the other has loaded; or companies by files as ordered: the front rank remaining as a reserve.—Should the battalion be formed only two deep; the front rank will remain kneeled, and the rear rank will fire by files.

---

7. *The front and rear faces of the square or oblong in march are encreased;* by repeatedly adding to their flanks, 4, divisions from the column sides which are thereby shortened, and oblique outwards to cover: they are decreased by the 4 outward divisions of the front and rear repeatedly becoming part of the flank sides which are thereby lengthened, and oblique inwards to cover.——Thus either advancing or retiring the whole may diminish to two sub-divisions in front, or if necessary, to a double file marched off from the center of the leading face.

Fig. 94.

8. *When the square or oblong forms in line on one of its sides,* or on any named company which is placed in a given direction.—Each other company will be WHEELED up more or less, till it stands with its inward flank

flank perpendicular to its point in the new line, to which the whole will MARCH and enter fucceffively, the outward companies taking care not to impede the inner ones, which muft form before them. According to the part of the battalion formed on, will this operation be more or lefs complex. — Or,—This may be done by the facing and filing of each divifion from its inward flank, to its point in the new line where it will form up.

---

9. *If from open column of march, it is neceffary to make front in oblong 3 deep to both flanks.*—The leading divifion *Halts*, the other divifions of the column will take half diftance and *Halt*: the half divifions will WHEEL outwards, *Halt*, and form an oblong clofed in the rear by the laft divifion. When column of march is to be refumed, the half divifions will WHEEL backwards into column, and the battalion will proceed.———If there are feveral battalions in the column; each will form as above, clofed by its own front and rear divifions; and the diftances betwixt battalions will alfo be clofed.

---

10. *If a battalion is marching in open ground, where it is neceffary to be prepared againft the attack of cavalry:* It may move in column of companies at quarter diftance, one named company in the center being ordered to keep an additional diftance of 2 files; in this fhape the battalion is eafily managed, or directed upon any point.———When the column HALTS, and is ordered to FORM THE SQUARE; the firft company falls back to the fecond; the laft company clofes up to the one before it: The whole companies make an interval of 2 paces in their center, by their fubdivifions taking each one pace to the flanks; 2 officers with

Fig. 95.

with their ferjeants place themfelves in each of the front and rear intervals, 2 officers with their ferjeants alfo take poft in rear of each flank of the company from which the additional interval has been kept, and a ferjeant takes the place of each flank front rank man of the firft divifion, and of each flank rear rank man of the laft divifion; all other officers, ferjeants, the 4 difplaced men, &c. &c. affemble in the center of the companies which are to form the flank faces.——Thofe laft named companies having been told off each in 4 fections, WHEEL up by fections, 2 to the right and 2 to the left; (the 2 rear companies at the fame time clofing up, and facing outwards) the inner fections then CLOSE forward to their front ones which drefs up with the extremities of the front and rear companies, and 4 files on each flank of the fecond companies from the front and from the rear, FACE outwards.——The whole thus ftand faced outwards and formed fix deep with 2 officers and their ferjeants in the middle of each face to command it; all the other officers as well as ferjeants, &c. are in the void fpace in the center, and the files of the officers in the faces may be compleated from ferjeants, &c. in the interior in fuch manner as the commandant may direct.—The mounted field officers muft pafs into the center of the column, by the rear face if neceffary opening from its center 2 paces and again clofing in.

Fig. 95.

When ordered, the 2 firft ranks all round the column will kneel and flope their bayonets, the 2 next ranks will fire ftanding, and all the others will remain in referve; the file coverers behind each officer of the fides will give back, and enable him to ftand in the 3d rank.——*When the march is to be refumed;* the fections that clofed up, fall back to their diftance, the fections then WHEEL back into column; the officers, ferjeants, &c. take their places on the flanks; and when the column is again put

in

in motion the companies that closed, successively take their proper distance.

Unless the companies are above 16 file they cannot be divided into 4 sections: If therefore they are under 16 file and told off in 3 sections, the column will march at the distance of a section, and in forming the square, the 2 outward sections will wheel up, but the 3d one will stand fast, and afterwards by dividing itself to right and left, will form a 4th rank to the others; in resuming column the outward sections wheel back and the rear of the center sections easily recover their places: as to all other circumstances they remain the same.

BY the foregoing Regulations, and the Rules they lay down, is every battalion; to direct its practice; to regulate its parades, guards, and field exercise; to disuse whatever is contrary and repugnant to them; and in no instance to deviate from the principles they contain, for to their strict observance is every one enjoined.——Among many other essential circumstances they pointedly require. Hurry and disunion to be avoided: order and mutual effort to be held sacred: ranks and files closed: music to be disused in instruction, march, or manœuvre: uniformity of position: equality of step in length and cadence: accurac of distances: precision of file marching: movements and formations made on determined points and lines, and mounted officers ready and accustomed to give such points and lines: alertness and intelligence in officers; energy and decision in their commands: modes of execution

tion fully determined and never varying.——Thus previous explanation being no longer neceſſary, prompt performance in all ſituations may immediately follow the ordered meaſures of the commander.

D. D.

End of THIRD PART.

INSPECTION or REVIEW

OF A

BATTALION of INFANTRY.

THE battalion marches to its ground in open column of companies or half companies—marches into the alignement by companies—forms in close order—takes open order as directed in the formation of the battalion.

In this disposition and the whole dressed to the right, the general is awaited.—He is to be received with the compliments due to his rank, as set forth in the regulation of military honours.—The colonel and lieutenant colonel on this occasion are on foot at the head of the colours; at all other times they are to remain on horseback.

A camp colour is to be originally placed 80 or 100 paces in front of the center of the battalion, where the general is supposed to take his station; but although he may chuse to quit that position, still the colour is to be considered as the point to work upon, and to which all movements, and formations are relative.

Receiving

Receiving the General.

PRESENT ARMS. { When the reviewing general prefents himfelf before the center, and is 50 or 60 paces diftant, he will be received with a general falute.—The men prefent arms, and the officers falute, fo as to drop their fwords with the laft motion of prefented arms: the mufic will play, and all the drums will beat:—The colours only falute fuch perfons as from their rank and by regulation are entitled to that honour.

SHOULDER ARMS. { The men fhoulder, and the officers recover their fwords with the laft motion.

The general then goes towards the right, the whole remaining perfectly fteady without paying any farther compliment while he paffes along the front of the battalion, and without facing when he goes along the flank and rear.——While the general is going round the battalion, the mufic will play, and the drums beat; they will ceafe as foon as the general has returned to the right flank of the battalion.

REAR RANKS TAKE CLOSE ORDER. MARCH. { While the general is proceeding to place himfelf in the front, this command will be given, and the colonel and lieutenant colonel will then mount on horfeback, in the rear of the center.

Marching paft in Ordinary Time.

COMPANIES ON YOUR LEFT, BACKWARD WHEEL.

MARCH.

Halt, drefs.

MARCH.

{ The battalion will break into column of companies the right in front.—The column is put in motion, pioneers and mufic having been previoufly ordered to the head of it.—Points will be afcertained by the adjutant for the exact, and feveral wheelings of the divifions, fo that their right flanks in marching paft, fhall be only 4 paces diftant from the camp colour, where it is fuppofed the general places himfelf to receive the falute.

The

The several companies wheel successively at the first angle of the ground.	*Halt, left, wheel.* *Halt, dress.* *March.*

The companies successively make this wheel, at the second angle of the ground, and which brings them on the line, on which they pass the general.—Each leader of a company when it has advanced 6 paces from the wheeling point, changes quickly by the rear to the right flank of his company, and as soon as he has placed himself on that flank, he will order eyes to be turned to the right.	*Halt, left, wheel.* *Halt, dress.* *March.* *Eyes, right.*

The leading company, and each other successively as it arrives within 50 paces of the general, opens its ranks, at which time the officers move into the front of the company, and the leading one is replaced on the right flank, by his serjeant.	*Rear ranks take open order.*

In *marching past* the reviewing general, the colonel is to be at the head of the grenadier company with the major a little behind him on his left.——The music are in two ranks 6 paces before the colonel: The pioneers are in two ranks 6 paces before the music, having a corporal at their head to lead them: The drummers and fifers are on the left flank of their respective companies.

The lieutenant colonel is to be in the rear, but in the absence of the colonel, the lieutenant colonel will of course supply his place.—The adjutant is in the rear, behind, and on the left of the lieutenant colonel.

The colours are 3 paces behind the 4th battalion company covered by their serjeants.——Staff officers do not march past.

In marching past at open ranks, the serjeant who is on the right flank of the company is responsible for the proper wheeling distance being kept from the

front

front rank of the company preceding him.——The leading officer muſt invariably preſerve his diſtance of 3 paces before the right of the company, and not derange its march, the rank of officers dreſs to him, eyes are turned a little to the right, and they divide the ground in order to cover the front of the company: if there is only one officer with the company, he is towards the right of it.——Supernumerary ſerjeants are 3 paces in the rear of their ſeveral diviſions.

The muſic begin to play, juſt after the leading company has made the ſecond wheel, they continue to march on, and do not draw up oppoſite the general.—They as well as the pioneers regulate their march by the head of the column.

The officers when they arrive at a proper diſtance from the general, muſt prepare to ſalute ſucceſſively by companies when within 6 paces of him, and recover their ſwords when 10 paces paſt him, without in the leaſt altering the rate of march, or impeding the front rank of companies.——The commanding officer when he has ſaluted at the head of the battalion, places himſelf near the general, and remains there till the rear has marched paſt.—The drummers give a roll, each when the officers of his own company ſalutes.

Rear ranks, take cloſe order. { The officers commanding companies will each ſucceſſively when he has paſſed the general by 30 paces, cloſe his rear ranks, and at this time each individual of the company reſumes the poſt which he held, when the column was firſt put in motion.

Halt, left wheel.
Halt, dreſs.
March.
{ The ſeveral companies wheel ſucceſſively when oppoſite the ground where the left of the regiment ſtood, their leading officers having ſhifted to their left flank, when the ranks cloſed.

HALT.
SUPPORT ARMS.
{ When the leading company is near to where the left of the battalion ſtood, the whole halt, muſic ceaſes, arms may be ſupported, and the quick march may inſtantly commence.

Marching

Marching paſt in Quick time.

The whole march off in quick time.——No muſic. | QUICK MARCH.

The column makes three ſeveral wheels, viz. at the point where the left of the battalion firſt ſtood: at the point where the firſt wheel was made: and at the point where the ſecond wheel was made, which places it on the line of paſſing the general. | *Halt, left, wheel.* *Halt, dreſs.* *Quick march.*

Before the leading company has made the laſt wheel arms are carried—When it has compleated that wheel, the muſic begin to play. | CARRY ARMS.

In *marching paſt* the general in quick time and at cloſe order, officers do not ſalute or pay any compliment, but are attentive to preſerve the proper intervals betwixt their companies.——The leading officer of each company ſhifts to its right by its rear in the ſame manner as in the ordinary march, 6 paces after the laſt wheel, which brings him on the line with the general, and when he has paſſed the general 30 paces he will reſume his proper pivot flank.———The ſupernumerary officers, and ſergeants march in a rank in rear of the companies at one pace from the rear rank, and officers ſwords are carried againſt the right ſhoulder, and ſteady.

The colonel, lieutenant colonel, major, and adjutant, are in the ſame places, as in marching paſt in ordinary time; as alſo drummers, pioneers, and muſic, which laſt will commence playing, juſt after they have wheeled into the line of paſſing, and will continue to march on at the head of the column.

The ſeveral companies 30 paces after paſſing will ſucceſſively dreſs to the left the proper pivot flank, and the officers will ſhift to that flank. | *Eyes, left.*

The

Halt, left, wheel. *Halt, dress.* *March.*	The companies succeffively wheel, when oppofite to the ground, where the left of the battalion flood.
HALT. MARCH.	When the head of the column approaches to the left of the ground on which it originally received the general, the mufic will ceafe, and the column will be halted in order to take up the ordinary march, for the purpofe of moving on an alignement.
Halt, left, wheel. *Halt, dress.* *March.*	When at the point on the left of the alignement.

Forming in Line.

HALT. LEFT, WHEEL INTO LINE. MARCH. *Halt, dress.*	The column prolongs the alignement, till arrived at the point where its head or right is to be placed.—It receives the word halt, pivots are inftantly corrected if neceffary, it wheels up into line, and the pioneers and mufic go to their pofts behind the center.
WITH CARTRIDGE: PRIME, AND LOAD.	The battalion being now formed at clofe order; the commanding officer will order it to prime and load with cartridge, and will proceed with *Movements* and *Manœuvres.*

But fhould the performance of the Manual, and Platoon Exercife be required.

The commanding officer after the line has formed, gives a CAUTION that the manual and platoon exercife will be performed, and goes to the rear of the battalion.—The major advances to the front of the battalion, OPEN RANKS; UNFIXES BAYONETS; SHOULDERS ARMS; makes the officers and colours TAKE THEIR POST OF EXERCISE in the rear by facing to the right; MARCHING through the

feveral

several intervals occupied by the serjeants, and when 3 paces beyond the rear rank, they halt, and then receive the word FRONT: The commanding officer, lieutenant colonel, adjutant, pioneers, music, supernumerary serjeants, drummers, fifers, are at their posts in the rear, as when the battalion is formed at close order.

Manual Exercise.

The major proceeds with the manual as directed by regulation, observing that the front rank only, comes down to the last position of the charge bayonets, the others remain ported.—The serjeants who preserve in the front rank the places of the platoon officers, remain there steady during the whole of the manual, except that they charge their pikes, at the same time as the bayonets.

Platoon Exercise.

The major closes rear ranks for the platoon exercise, and platoon officers, and serjeants, and colours, and every other individual, take their places, as when the battalion is at close order.

The major proceeds with the platoon exercise; and the several ranks make ready each according to its situation of front, center, and rear; after firing, they load and shoulder agreeable to the regulation.

The manual and platoon exercise being finished, the major goes to his post, and the commanding officer of the battalion, proceeds to PRIME AND LOAD! with cartridge, and then to commence the ordered movements.

Movements.

Movements.

Plate 16.

		Sect.	
1. *On a rear division.*	Form close column of companies behind grenadiers	137	The column marches quick 20, or 30 paces to the right, and without halting begins to deploy into line on the rear division.—The commanding officer of the battalion gives the word for each division to halt, front.
	Form close column of two companies	147	
	Face and march to the right	150	
	Deploy on the rear division	149	
2. *On a front division.*	Form close column of companies, in front of the left	137	The column marches quick 30 or 40 paces to the left, and without halting begins to deploy on the front division.—The commanding officer of the battalion gives the word for each division to halt, front.
	Form close column of two companies	147	
	Face and march to the left	150	
	Deploy on the front division	148	
3. *On a central division.*	Form close column of companies, on a central company, either flank in front, and facing to the rear	138	The close column is formed facing to the rear.—It then countermarches each division so as to return to the proper front.—In the central deployment by companies, the company officers give the words to halt, front.
	Countermarch of each division in close column	143	
	Deploy on any central named company	146	
4. *Change of position in open column.*	Wheel back into open column of companies, the right in front	108	
	March forward 30, or 40 paces	111	
	Enter an oblique line (the 3 or 4 leading companies) by wheeling successively to the left, a half wheel,	126	The battalion thus, at an intermediate point enters an alignement on which it is to form.
	Halt.		
	The rear companies file into column	126	
	Wheel up into line	118	

The

			Sect.
5. **Wing thrown back.**	The left company is wheeled back, till parallel to the original position.—The rest of the companies wheel into echellon.—March to the rear.—Form on the left company.	160	The whole companies wheel back at the same time; the left company twice the number of paces that the others do.—Should it be necessary for the subsequent movements the line may retire 50, or 60 paces, and then front.
6. **Countermarch and change of position.**	Wheel back into open column, the right in front — —	108	
	Countermarch companies by files	100	After the countermarch by files, the column stands with its left in front.—The column closes in quick time.—The square is formed, and close column reformed as in part 4th, S. 189.—The column opens out in quick time from its rear division and halts.—The countermarch by companies from the rear to the front is in ordinary time.—When the line is formed, it is then considerably to the general's right, and with its rear to him.
	March in column, 30 or 40 paces—Head division halts close to the head of column	138	
	Form square, and prepare for firing Reform in close column —	189	
	Open out to open column from the rear, and halt — —	153	
	Change head of column, by the countermarch of companies, from the rear to the front — —	101	
	Column moves on, and halts		
	Wheel up into line		
7.	Countermarch by files, on the center of the battalion — —	98	This brings back the battalion to its original front.
8. **March in open column.**	Form open column, behind the left company which is put in march when the 3d company has taken its place in column	121 / 2 / 121	The companies that are filing incline towards the head of the column: successively front at their wheeling distances ascertained as usual by their serjeants: take up the ordinary step, and follow in open column.
	The right subdivisions double —	87	
	The right subdivisions move up —	88	
	The column halts, and pivots are corrected.		When the column is marching steadily, the whole sub-divisions double at once by one command, and again move up at another.
	Wheel up into line.		

When

[10]

		Sect.	
9. *Echellon change of position.*	Wheel back into open column the left in front	109	The line is thus formed oblique from open column, on a central company, by the echellon march.
	The third company is wheeled back, the 8th of the circle, and each of the others 3-16ths of the circle	158	
	Form line on the third company by the echellon march	159	
10. *Echellon change of position.*	The left company is wheeled up the 8th of the circle, and each of the others 1-16th —Form line by the echellon march	159	The line thus changes position to the front, on the left company, by the echellon march.
11. *Change of position.*	The battalion faces to the right—Marches in file (50, or 60 paces). Forms column of companies, on the march — Halts—Wheels up into line, except the light company, which files quickly to the right, and forms behind the colours	94 95	The column of companies is formed by the rear men of each moving up quick to the left of their leaders, and of each other: the officers move to pivot flanks, and pivots are instantly corrected.—The column halts when the colours are opposite to the general.
12. *Retreat in line.*	The battalion retires (50 paces)—Halts fronts—Fires twice by companies from center to flanks	168	The light company being previously subdivided and prepared, acts in the retreat by alternate companies as directed in S. 176. and when the line halts and fronts, it resumes its place on the left.
	Retire by alternate companies in two lines, (250 paces) each retreat about 50 paces	176	
	Form line.		
	Retire in line (50 paces)—Halt, front.		

Companies

[11]

13.

March to a flank in echellon.

	Sect.
Companies make a half wheel to the right	154
March in echellon (250 paces) —	155
Wheel back on the march into parallel line	156
Forward (100 paces)—Halt.	
Fire thrice by companies from flanks to center.	

At the word wheel back into line (the pivot flanks mark time and the divisions wheel back in ordinary time.—At the proper instant when the battalion is formed, the commander gives his word forward, for the whole to advance by the colours, and to correct any irregularity that there may be in the battalion.

If the battalion has hitherto been formed two deep, it will now form three deep if its companies are of ten files each.

14.

Movements in the square.

Form square.
March the square by the left angle of the front face (50 paces)—Halt—Form square.
March square by the left face—Halt—Form square.
March square by the rear face (60 paces)—Halt—Form square.
Fire in square by companies.
Form the line.

178 {

The square is formed by the echellon march of companies.

After the march by the left face, the square is formed when it is opposite to the general.

The firings in square are as expressed in S. 178.

The line is formed by the echellon wheel up, and march of companies

When the order is given to form line, the light company marches quickly, and places itself two deep and in two divisions 10, or 12 paces behind the two center companies.

15.

Retiring, and filing to the rear.

Retire in line (100 paces) —	168
File by companies from the proper right— Halt in open column the right in front	175
Wheel up into line.	

When the line has passed the light company 20 paces, that company extends to cover the center of the battalion, and follows at 50 or 60 paces distant; and when the column halts to form, the light company passes quickly through and beyond it.

The companies file quick to the rear.

The battalion forms line at the extremity of its ground; the light company 30 paces in its rear.

Advance

16.

Firing—Advancing—and charging to the front.

	Sect.
Advance in line 50 paces.	166
File from the right of companies to the front (50 paces)—Halt in open column the left in front—Wheel up into line.	178
Advance in line (50 paces)	
Advance by alternate half battalions, and fire four times	177
Form line—Advance (50 paces)—Fire volley.	
Advance (20 paces)—Fire volley—Charge bayonets (50 paces)—Halt.—Load.	

Before the line advances, the light company quickly forms extended 30 paces before the center, and preserves that distance in advancing.

When the column halts to form, the light company passes quick to the rear, and assembles, half of it behind each flank, and moves relatively with the flank companies till after the charge of bayonets.

The alternate half battalions fire the 2 first ranks standing.

After the volley, bayonets are ported, the battalion advances firm by the center at the quick step, and at the word Halt, the front rank comes down to the charging position:—The word Prime and Load is then given, and the light company issuing from behind the flanks, pursue, return, and assemble and join on the left of the battalion.

17.

Retiring in line.

Retire in line (100 paces)	
Retire by alternate half battalions—Fire four times	177
Retire in line, 100 paces or more—Halt, front.	

The whole battalion being assembled.
The alternate half battalions, fire the two front ranks standing.

18.

Advancing in line.

Advance in line (100 paces) Halt—Fire twice, oblique to right and left	166
Advance in line (100 paces)—Halt—Fire two volleys—Port arms at the last one, and half cock	
Open ranks—Advance within 50 paces—Halt—General salute	

In the oblique firing, and in the volleys the front rank kneels.

The music may occasionally play, and drums roll, while the line advances.

The music will play, when advancing at open ranks.

Such other *Manœuvres*, as may at the time be required.

THE

THE number of paces mentioned in the several movements are not positively prescribed, but are supposed to be nearly such as will give the intended relative situations.—If the ground allows the marches to the rear and front to be longer, it will be so much the better.

No improper pauses should be made betwixt the connected parts of the same movement.—The detached points necessary in formation should be timeously prepared and given.

The advance of the battalion should instantly succeed the forming of the line; and when it arrives and halts at the point where it is to fire, the firing ought instantly to commence at the word Halt; for the battalion having been apprized during the march of the nature of the required firing, no improper delay need therefore be made.

The greatest care is to be taken by the officers and under officers in the rear (whose principal attention this is) that the rear ranks are well locked up in the firings, and that in loading they do not fall back.

The line if retiring, halts fronts, at one command; and instantly begins firing, having been apprized during its movement of the nature of the firing.

The pause betwixt each of the firing words—*Make ready! Present! Fire!* is the same as the ordinary time, viz. the 75th part of a minute, and no other pause is to be made betwixt the words.

In firing by *Companies by wings*: Each wing carries on its fire independent, and without regard to the other wing, whether it fires from the center to the flanks, or from the flanks to the center.—If there are five companies in the wing two pauses will be made betwixt the *fire* of each, and the *make ready* of the succeeding one.—

If

If there are four companies in the wing, three pauses will be made betwixt the *fire* of each, and the *make ready* of the succeeding one.—This will allow sufficient time for the first company to have again loaded, and shouldered at the time the last company fires, and will establish proper intervals between each.

In firing by *Grand Divisions*, three pauses will be made betwixt the *fire* of each division, and the *make ready* of the succeeding one.

In firing by *Wings*—one wing will make ready the instant the other is shouldering—The commanding officer of the battalion fires the wings.

In firing companies by *Files*: each company fires independent.—When the right file presents, the next makes ready, and so on.—After the first fire, each man as he loads, comes to a recover, and the file again fires without waiting for any other; the rear rank men are to have their eyes on their front rank men and be guided by, and present with them.

In general after the march in front, and halt of the battalion, company or platoon firing should begin from the center, and not from the flanks.—In other cases, and in successive formations it may begin from whatever division first arrives, and halts on its ground.

The Intention of fixing upon some of the most essential infantry movements, and thus ordering them to be executed by each battalion when seen separately, is—That thereby the *Inspecting General* may be enabled to report the more minutely and comparatively, on the performance by each battalion, of the great leading points of movement.

He

He will therefore among other circumstances, particularly observe and specify—
Whether or not,

The original formation of the battalion is according to order.

The marches are made with accuracy, at the required times and length of step, and on such objects as are given.

The proper distances in column and echellon, are at all times preserved.

The wheelings are made just, and in the manner prescribed.

The formations into line are made true, without false openings, or necessity of correction.

The officers are alert in their changes of situation, exact in their own personal movements, and loud, decided, and pointed in their words of command.

The march in line is uniformly steady, without floating, opening, or closing.

The march in file, close, firm, and without lengthening out.

The officers and under officers give the aids required of them, with due quickness, and precision.

Hurry and unnecessary delay in the movements are equally avoided.

In the firings, the loading is quick, the levelling just, the officers animated and exact in their commands.

When two or more battalions are inspected, or exercised together they will be formed in one line with the ordered interval.—They will receive the *General*; march past; and may perform the same identical movements as are before prescribed for the single battalion; observing the additional directions that are given for those of the line.

When the line of two or more battalions is *Marching past* in column of companies; it must occupy no greater extent of ground than when it originally wheeled into column.—The order is never to be broken, or lengthened out.—No particular battalion, or the artillery are allowed to encreafe diftances for their own partial appearance.—The battalion guns will march two a-breaft.—Ranks are one pace afunder, or if ordered to be open, the diftances between companies and battalions will not be encreafed.—The mufic of each battalion in paffing may play, but will continue to march on.—The ordinary march is preferved.—Officers do not falute marching, but when particularly ordered.

When a confiderable body of infantry, or when infantry and cavalry are united, and to act in *corps*; their combined operations, fuch as movements in columns, echellons, or lines; their formations; the conduct of attacks, and retreats, &c. depending on numbers, and circumftances of ground, or fituation; can only be determined and applied according to the views of the commander; but the great principles of movement laid down for the line will ftill direct, and the detail of execution will remain invariable, being compounded of thofe prefcribed for the *Company*, *Battalion*, and the *Line*.

LIGHT INFANTRY.

GENERAL ATTENTIONS.

WHEN the LIGHT INFANTRY companies are in line with their battalions they are to form and act in every respect as a company of the battalion, but when not in line they may loosen their files to six inches. *Distances of files*

Open order is to be two feet between each file the necessity of encreasing this distance must depend on circumstances and be regulated at the moment by the commanding officer. *Open order.*

The files may be extended from right, left, or center, according to circumstances—in executing it each front rank man must carefully take his distance from the man next to him on that side from which the extension is made—the rear rank men conform to the movement of their file leaders. *Manner of extending.*

When the company is not in extended order—all firing is to be by single men, each firing as quick as he can, consistent with loading properly—the firing to begin from the flank, or from the point first formed.

[2]

Firing.

In firing in extended order, it is to be a standing rule that the two men of the same file are never unloaded together, for which purpose as soon as the front rank man has fired he is to slip round the left of the rear rank man, who will make a short pace forward and put himself in the others place whom he is to protect while loading—when the first man returns his ramrod he will give his comrade the word *ready* after which and not before he may fire and immediately change places as before.

Advancing and retreating.

To cease firing.

The same method of firing to be observed when advancing or retreating which must always be in *ordinary time* (especially if cannon are ordered to the front with the light companies which may often be the case).—Particular attention must be paid to cease firing on the first word or signal for that purpose.

Movements in quick time.

All movements of the light companies except when firing, advancing, or retreating are to be in *quick time*.

Never to run unless ordered.

The light companies are never to run unless particularly directed, and in that case they are only to run at that pace in which they can preserve their order—and it is to be a rule that the two men of the same file never separate on any account whatever.

Avoid confusion.

The utmost care to be taken to avoid confusion which too much hurry, even in the smallest bodies will certainly occasion.—The intermixture of files can never be allowed of.

File movements.

Though all movements should be made in *front* as much as possible yet from the nature of those of light infantry and the ground they are more particularly liable to traverse, file movements may frequently be necessary.——All such to be made from one of the flanks by previously facing to it, and the files to loosen so as to march perfectly at ease but not more.

In

In forming, the inversion of files or of ranks is not to be attended to if time is thereby gained.—*Forming to the front* to be done by the file moving briskly up to the right or left of the leading file as ordered. Forming to the front.

Forming to right or *left*—the leading file will halt and face as directed, as will the succeeding ones as they come up to their proper distances. Right or left.

Forming forward to right or *left*—The leading file halts and faces as directed—the succeeding files lead round the rear and form to the same front as the leading file has done, and at their proper distances. Forward to right or left.

When marching to the rear by files and to *form to the front*—The leading file will halt and front, the succeeding files will go round the rear of the leading file and form on the right or left of it as directed.—*Forming to right* or *left*—or *forward to right* or *left* is done in the same manner as when marching to the front. Marching to rear and forming.

All *signals*—*words of command* and *directions* are for the officer commanding the company or division who gives the necessary orders in consequence. Signals, &c. for officers commanding.

The *necessary signals* will be previously settled, and as they will be very few and simple, the officers and non-commissioned officers are expected to be masters of them. Signals.

The officer commanding the company will be on the right covered by a serjeant.—The next on the left also covered by a serjeant.—The youngest officer in the rear.—In extended order the post of the officers and serjeants is always in the rear equally divided, where they must pay particular attention that the men preserve their order, and that they level, fire, and load cooly and properly, they must likewise be attentive to direct them to the supposed object of attack. Post of officers.

In

In marching by files the officer commanding leads—by divisions each officer leads one.—The supernumerary officer, if there be one, is in both cases with the officer commanding, ready to obey any directions he may receive from him.

Taking post.   When a light company, or detachment, is ordered to take post on any particular spot, it is to be the business of the officer commanding it, to take the best advantage of the ground, observing that he must never disperse his company—but if it should be necessary to make small detachments from it, he must still preserve a part of his company or detachment as a reserve, on which those detachments may fall back—and this is to be a general rule in all cases where the strength of the party is sufficient to allow of making detachments from it.——

To cover in situations of defence.   The officers must also see that in situations of defence the men cover themselves with trees, walls, large stones, or whatever may present itself.—In firing from behind trees, large stones, &c. they are to present to the right of the object which covers them, and in changing places with the other man of the file after firing, they will step back and to the left, so that the rear rank man may step forward without being exposed.

Arms how carried.   The arms of light infantry in general will be carried sloped, and with their bayonets fixed.—Flanking and advanced parties however, or parties in particular situations, may carry them trailed, and without bayonets, for the purpose of taking cooler and more deliberate aim.

LIGHT

[5]

LIGHT INFANTRY ATTACHED to respective regiments when in LINE.

The light company will be posted in the rear of its respective regiment, divided in two divisions, that on the right will be in the rear of the second company; that on the left, in the rear of the seventh company, and they will at all times observe the distance of thirty paces.—The captain, or officer commanding, will be with the right division. Divisions cover 2d and 7th companies.
Post of commanding officer.

When the line breaks into column, if the light companies receive no particular directions for covering either the front or flanks of the column, they will wheel as the companies of the battalion do, and conform themselves exactly to the movements of the second and seventh companies, so as at all times to be in their proper places. Line breaks into column.

If the line forms a close column, and the light companies receive no particular directions, they are to form by companies, and close up in the rear of the column, in the same manner as their respective battalions. Line forms close column.

When the column deploys into line, the light companies will face each as its battalion does, file with it in the rear; and when the battalion forms in the line, will take its proper post, in divisions behind the second and seventh companies. Line deploys.

If the light companies are ordered to cover the line to the front, either by word or signal, the divisions will move to the front from their inner flanks, round the

flanks

flanks of the battalions, and when at the diſtance of fifty paces, the leading flanks will wheel towards each other, ſo as to meet oppoſite the center of the battalion; opening their files gradually from the rear, ſo as to cover the whole extent of the battalion—the ſerjeant-coverer of each diviſion attending to the files taking their proper diſtance—the files are to halt and front of themſelves—In this poſition, and in all extended order, the poſt of the officer commanding is in the rear of the center, and the movements are to be regulated by the company belonging to the battalion, which regulates thoſe of the line.

Cover front of Battalions.

Poſt of commanding officer,

When the light companies are called in, the line may either be halted or advancing—In the firſt caſe, they will retire towards the line, cloſing to their outer flanks by degrees, ſo as when they come near their battalions, they may be in two diviſions, ready to file round the flanks of the battalion to their places.—If the line is advancing, they will only cloſe to their outer flanks, ſo as to be in two diviſions by the time the line comes up to them, when they will inſtantly face outward, and file to the rear.

Line halted or advancing when light infantry are called in.

LIGHT INFANTRY Companies formed in BATTALION.

When the light infantry companies are aſſembled in battalion, their movements muſt be on the ſame principles as thoſe of the line; the officers and non-commiſſioned officers poſted in the ſame manner, and as far as poſſible the ſame words of command ſhould be uſed; it is in their rapidity alone that they muſt be diſtinguiſhed,

Movement ſame as the line.

ed, to facilitate which, the files are to be loosened to the distance of six inches, but great care is to be taken that rapidity does not degenerate into confusion.

When two or more companies are together, they are to consider themselves as a battalion, the senior officer is to take the command, leaving the immediate command of his own company to the next officer belonging to it.——As Light Infantry seldom act in large bodies, all their movements may be in quick time; but, when in column, the same attention must be paid to the pivots covering, and the preservation of distances, as is done by the line—the doing so, will always be found the quickest way of forming, by precluding the necessity of much after dressing.

Quick time.

Covering pivots, &c.

In marching in line to the front, a regulating company must be named, by which the others must carefully dress, and whose movements they must follow.—The officer leading this regulating company, must take points on which to march perpendicular to the front of the battalion, and must lead steadily on them, though in quick time — without these precautions, and great attention being paid to them, the march in front must soon become irregular, the files will inevitably intermix, and great confusion must be the consequence.

Regulating company.

A battalion of light infantry may occasionally be ordered to run, for the purpose of anticipating an enemy going to occupy any particular post;—but in doing so, the utmost care is to be taken that confusion do not ensue; for which purpose the velocity must never exceed that at which the divisions can keep together and dressed; the distances must be preserved as much as possible.—Running must generally be in a column; but in a case of absolute necessity to make a very quick movement to the front, with a battalion of four or five companies or more, the best, and easiest way of doing it without confusion, will be in *echellon*, by companies, each retired six paces from the preceding one.

May occasionally run.

But generally in column.

May in echellon.

A ll

All Columns of light infantry to be formed by sub-divisions that is half companies.

Forming from open column.

The forming from open column to the front may frequently be done by the divisions obliquing to the right or left of the leading division, and if necessary firing as they come up.——Light infantry firing in divisions is to be always by single men as directed in general attentions.

Firing in divisions.

Movement by files.

Battalions of light infantry may frequently find it necessary to move by file through woods and over very rough countries---in all cases where it is practicable it is to be done from the right or left of companies, and distances must be preserved for forming in the quickest manner possible---whenever one company *forms* the rest are to do the same even supposing they do not hear the word or signal for that purpose.

Forming in front.

If to *form to the front* the leading files of each company halts and dresses, the rest move up to the right or left of them to their proper places.

Forming in right or left.

If to *form to the right* or *left*, the companies first form separately and move up and dress with what will then be the front company, by which means the officer commanding will have it in his power to keep such companies in reserve as he thinks proper, as also in forming to throw them to the right or left of the front company as circumstances may require---the companies which are to dress with the front company are to move up to it obliquely in line.

Advanced and flanking parties.

A battalion of light infantry marching through a wood should have parties in front and on its flank in proportion to the strength of the battalion---the parties should march in front with extended file, and if attacked must take post and defend themselves till supported or called in.

When

[9]

When ordered to secure a wood of no very great extent the battalion should go through it, and take post on the opposite side within its skirt, so as to have the plain before it, in this as well as in all other cases, parties should be detached 30 or 40 yards on the flanks. To secure a wood.

When firing in line advancing, the march must be very slow, the line must be preserved, and the officers must take care to point out the supposed object of attack, and see that the men direct their fire to it—very particular attention is to be paid that the fire is directed to the proper object, and that it ceases on the first word or signal for that purpose. Firing in line.

When the light infantry in battalion is detached from the line, the officer commanding must take care to understand thoroughly, the nature of the intended movement, so as to be certain of co-operating with the line with exactness and precision. Co-operation with line.

In general the method of taking post with a battalion of light infantry, whether large or small, must depend upon the intelligence of the officer who commands it, but he must observe the same rule as was given for a company, viz. Whatever detachments he may find necessary to make always to keep the most considerable part together as a reserve. To take post.

The success of any engagement in a wood or strong country depends upon the coolness and presence of mind of the commanding officer and the silence and obedience of the men, fully as much as upon their bravery. Commanding officer.

The arms of the light infantry when in battalion, while in movement are generally to be sloped, but always by order and their bayonets are to be fixed. Arms how carried.

If at any time a battalion of light infantry is ordered into the line, the files must be closed, and it must in every respect act as other battalions of the line. Light infantry in line.

The Signals.

The SIGNALS—To *Advance* : To *Retreat* : To *Halt* ; To *ceafe* firing : To *affemble*, or call in all parties : are to be always confidered as fixed and determined ones, and are never to be changed.—The bugle horn of each company is to make himfelf perfect mafter of them.

All fignals are to be repeated.

All of thofe fignals made from the line or column are to convey the intention of the commanding officer of the line to the officer commanding the light infantry who will either communicate them to the feveral companies or detachments by word or fignal.

PART the FOURTH.

OF THE LINE

All great bodies of troops are formed in one or more lines.

>Each line is divided into right and left wings.
>Each wing is composed of one, two, or more divisions.
>Each division is composed of one, or more brigades.
>Each brigade is formed of two, three, or four battalions.

These bodies have their immediate commanders, subordinate to each other.

Battalions are formed in line at a distance of 12 paces from each other, and this interval is occupied by 2 cannon, which are attached to each battalion.——There is no encreased distance betwixt brigades, unless particular circumstances require it.—In exercise should there be no cannon betwixt the battalions, the interval may be reduced to 6 paces.

Movements of a Line.

1. THE movements and manœuvres of a confiderable line are fimilar to, and derived from the fame general principles as thofe of the fingle battalion; they will be compounded, varied, and applied according to circumftances, ground, and the intentions of the commanding officer; but their modes of execution remain unchangeable, and known to all.—The greater the body, the fewer and the more fimple ought to be the manœuvres required of it.

Circulation of commands.

2. If feveral regiments exercife or manœuvre together, the commanding officer of the line or column gives his fhort orders of caution or execution to the commander of the regulating battalion of the line, or of the head of the column, where he himfelf generally is; and fometimes to the commander of the battalion to which he is then neareft, and each battalion commander repeats them loud without delay.—— When any complicated or combined movement is to be made, which requires previous explanation; it muft be communicated clearly to the commanders of corps by detached officers, before its execution can be ordered to commence.—The feveral chiefs of brigades, &c. watch over, and direct the interior movements of their refpective bodies; they repeat the general orders of execution given, if they fee that it is neceffary; and announce fuch preparatory ones as are verbally fent to them.

When the general order is not heard or underftood by part of a line, each battalion commander (where the intention is obvious) will conform

form as quickly as possible to the movements which he sees executed to his right or left, according to the point from whence the movement begins; but platoon officers execute only, on the orders of their battalion commander.

4. The commander in chief, will always himself loudly announce his commands of execution as MARCH, or HALT, and the commanders of battalions will without waiting for each other, endeavour in the same moment to repeat them: If officers are quick, observing, firm, and decided in their commands, such repetition will be instantaneous. *Commands of execution.*

5 It is impossible to ascertain the words of command to be given in all cases.—Where such are not pointed out, they must depend on the circumstances of the situation, and be short, clear, and expressive of what is to be done.—Where they are not comprehended, they must be repeated and no operation begun, till its intention is well understood; otherwise that disorder, which may be originally prevented, is not easily remedied if once it has taken place.

6. When troops are halted, explanatory cautions are proper before they are put in march; but when they are in motion, and in situations where perfect correctness is expected, as in the march in line, and in the prolonging of an alignement, no caution should precede the word HALT, but the whole should at once firmly halt.—This is to be understood of a column of manœuvre, but where a column of march is unavoidably from impediments of the route a little opened out, and that its head stops in order to remedy such extension, or to form in line, the several rear battalions will be halted successively at their just distances. *Cautionary commands.*

Regulating body in movement.

7. The movements of all great bodies are made either in line or column.—In line they are in general regulated by a battalion of that flank which is nearest to, and is to preserve the appui, or which is to make the attack: In column, they are directed by its head, and the commander of the whole is with the regulating body: There are very few cases in which the center ought to regulate, although the direct march of the line in front, appears to be the easiest conducted by a battalion of the center.—If an enemy is to be turned or an attack made, it is by the flank that such movements are led: It is the flank that must preserve the line of appui in all movements in front: If the line is thrown backward or forward, it is generally on a flank point: If the line breaks into column, it is the head or leading flank of that column which conducts, and whose writhes and turnings are followed by every other part of the body, and such head becomes a flank when formed into line: It is seldom that an attack is formed from the center and a movement seldomer need be.—The commander will therefore be on which ever flank directs the operations of the line, and by which he proposes to make the attack, or to counteract the attempts of the enemy.

Reserves.

8. No considerable body should ever be assembled or formed for action, without a proportion of it being placed in reserve or second line, and more or less strong according to circumstances.

Supporting lines.

9. Where several and supporting lines of attack are formed, the second should out-flank the first; the third the second, &c. the advanced one being thereby strengthened and supported on its outward wing.

Cannon.

10. The cannon attached to battalions whether in line or column, will

will accompany in all situations the movements of their proper battalions.—Those that are brigaded make a separate object.

11. The general firings of the line, are executed separately, and independantly by each battalion.

Firings.

12. The chief commander of a line must have several mounted officers, or other intelligent persons at his disposal, both to circulate his orders, and to mark, and determine such original points as become necessary in movement.—The adjutants of battalions are in general wanted to assist in the separate formations of their battalions.

Necessary officers.

13. Partial signals of the drum for a battalion, must not be given in line.—But from the battalion where the chief commander is and by his particular direction, such signal may when proper be made for the whole (but not repeated): If halted and standing at ease to assemble: If assembled to be ready to march: If firing in line for a general cessation: and before a march to mark the proper cadence by 5 or 6 strong taps.—Signals that cannot with propriety be applied in service, should not be used in exercise; and it is evident, that no loud signals or even commands, or music or drums, can be used in columns of route or in movements made near to, though not in presence of an enemy, as it is most important on such occasions to conceal them, and not unnecessarily to discover them to the enemy.

Signals of the drum.

14. Although in general the INVERSION of all bodies in line is to be avoided; yet there are situations where this rule must be dispensed with, and the quickest formation to a particular front, thereby obtained.—The battalion or line may be obliged to face to the right about

Inversion of the line in formation, sometimes necessary.

the

the more readily to oppose a danger, instead of changing its position by a countermarch: it may even be under the necessity of forming to a flank, with its rear rank in front.—The column with its right in front may arrive on the left of its ground, and be obliged immediately to form up and support that point, so that the right of the line will become the left.—Part of a second line may double round on the extremity of a first line, thereby to out-flank an enemy.—A corps moving to a flank by lines, may be obliged in the quickest manner to form up to the front of its march, so that the new lines shall be composed each of parts of the old ones.—Many other situations may be imagined, where opposing the rear rank admits of no choice, and where an inversion of the divisions of the line, will gain much time, and becomes absolutely necessary when the formation is required from the point of appui, and near to an enemy.—Troops must therefore be accustomed to such operations; but the application of them requires great method and recollection, otherwise in such critical situations confusion is very easily produced, and will even be attended with the most fatal consequences.

Open Column of the Line.

1. The great changes of situation of the line are performed in open column of manoeuvre.

2. The line breaks into open column by wheels of the quarter circle.

3. The

3. The general circumstances attending the open column have been already explained under that head, part the 3d.

4. The several general directions given for the single battalion in open column, extend to each battalion that makes part of a considerable column, and their minute observance is then most especially essential, and must be carefully recollected.

5. In open column, the leading division of each battalion, will preserve the distance of intervals betwixt battalions, in addition to that of its own front.—The column of companies or sub-divisions marching at half or quarter distance will preserve an interval between battalions equal to the front of the column. Intervals.

6. Battalions are to a line, what companies are to a battalion.—Not only the whole divisions of a battalion, but the whole battalions of a line or column should MARCH off, and HALT together; and to ensure this in the exercise of considerable bodies, signals of cannon are often given for such purpose. In most situations the quick circulation of verbal commands must be sufficient.

7. The same rules that direct the entry and march of one battalion in an alignement on which it is to form, (S. 115. 118.) apply with encreased attention to those of the most considerable column.—The point where the head of the column enters an alignement, and which is never quitted by a mounted or other officer but as he is relieved, and until the whole have entered; the point where the head of the leading battalion halts, in order to form; the several adjutants who place themselves in the true line; the prolongation of battalions which may Points of formation in open column.

have

have formed up; all thefe are fo many marked points within the line itfelf, on which the dreffing of pivots or battalions can be regulated, either while marching in the line, or when each halts and is to be corrected, in order to wheel up into line.

<small>Poft of commanding officer and adjutant in formations.</small>

8. When the head of a battalion in a general column, or in its undividual column, halts on a line on which it is to form; the commanding officer muft invariably be at that head point, inftantly to correct his pivots on the adjutant, who is invariably at the rear of the battalion in the true prolonged line on which it is marching or which it is to take up.—In like manner the commanding officer is with the leading divifion of an Echellon coming into line and the adjutant marks the other flank of the battalion on which the divifions are fucceffively corrected.

<small>Diftant points are of great advantage.</small>

9. In formations or changes of direction the commander in chief will if poffible preferve and procure confpicuous diftant points in their prolongation, which when known will affift others as well as himfelf, in keeping the line in the pofition he intends.

<small>Situations in which movements in open column are effential.</small>

10. The movements in open column of manœuvre are particularly neceffary.—When a line formed in order of battle is to extend in the fame direction to either flank in order to follow the march of an enemy, or to out-flank him if he remains pofted; nor is any movement more important, or can be more fecurely or effectually practifed againft an enemy inaccurate and inferior in difcipline, who in attempting the counter-movement is generally thrown into confufion.—Or, when arriving in column of march on any ground, the commander in determining the general direction that his line is to take, fhall not have been

able

able to ascertain the points where he would fix the flanks of it; but after entering into it, is obliged in consequence of the position or manœuvres of the enemy, either to stop his own movement sooner than he intended, or to prolong it, beyond the point he originally meant.

General Changes of Position of a Line.

Changes of Position of a Line composed of several battalions are according to circumstances effected by the *Echellon* march, the *filing* of divisions, or the *march* of battalion in open column,—and points in the new line will always in due time be ascertained, at which the leading division of each battalion is to enter.

1st. *When a considerable line is to take up a new position* Parallel *or nearly so to the old one, in front or rear of it, and facing either to, or from the old line.*   Fig. 96. E. D.

It may be done (according as the new line out-flanks the old line, connected with other circumstances)—by the march in *line*: the march in *Echellon* divisions: the *filing* of platoons.   } *If in front of and facing as the old line.*

Or, The line will break into open column to whatever hand the new position out-flanks the old one.—The several battalions are then disengaged and put in march in separate columns; flank points of entry

try for each are in the mean time preparing by the detached adjutants.—The leader of the 2d battalion from the directing flank has a point in or before the new line afcertained to him his adjutant not being yet fixed, and marches upon it: The leader of the firft battalion will preferve the parallelifm, or give gradually the new inclination to the heads of the other battalions.—Thefe during the march never having overpaffed the line of their leading ones, nearly dreffing up, and preferving their battalion diftances, arrive at their adjutants and form in line by wheeling, filing, or Echellon marching, as may have been ordered.

Fig. 97.

If in front of, and facing to the old line.
{ The battalion columns will as before enter and form on the line.— Within themfelves they need not be inverted, but the right of the line will now be the left, nor can it well be avoided, unlefs—by countermarching the line before the movement: or, by a complicated operation during the movement: or by countermarching firft the battalion, and then the line, after the movement.

If in rear of, and facing as the old line.
{ The fame identical operations according to circumftances are applied as when the pofition in front is taken, *facing as the old line.*—The line or echellons after facing about; or the heads of battalion columns after breaking; *march to the rear,* and front or form in line *facing as the old line.*

If in the rear of, and facing from the old line.
{ The battalion columns will lead to the rear and enter and form on the new line, the other circumftances will take place as when the new line is in front of and *faces to the old line.*—

The

The Echellon movements will not apply in this case without inverting the ranks.

2d. *When a considerable line is to take up a new position which (or whose prolongation* INTERSECTS *to the right or left of the old line, and which faces either to or from the old line.*   Fig. 96. C. B.

The line will break to which ever flank is nearest to the new position.—The heads of battalion columns will be separately conducted to their points in the new line, being regulated by the leading flank battalions; they will again enter into the general open column, and form in line by wheeling up.   *When the new line faces from the old line.*

Fig. 97.

Or the line after breaking to the flank, may continue its march in column, enter and prolong the new line, 'till its head halts at its point in that line.—The divisions of the leading battalion or of such other as then ought, will *file;* and the other rear battalions will disengage their heads, and separately march off in column to their several points of entry on the new line, which are marked by their adjutants.   Fig. 104.

If the angle formed by the two lines is not above the half of a right one, and that the flank of the new line is not very distant, this change may be made by the Echellon march of divisions.

The line will break into open column towards the new position.

When the new line faces to the old line. { fition.—The general column will enter the new line at its nearest point, prolong it if neceſſary, and when the head halts, the rear battalions will diſengage and march to their points of entry on the new line.

———

Fig 96. F.

3d. *When a conſiderable line has to take up a new poſition, which (or the prolongation of which) INTERSECTS the body of the line, and which faces to, or from the old line.*

When the new line interſects the body of the old line, and faces to either flank. } The poſition will be changed by the Echellon march of diviſions on the central point.

Fig. 45. 48.

Or, The diviſion which is in the point of interſection will place its pivot flank perpendicular to the new direction, and the line will break inwards and backwards facing to that diviſion.—The diviſions of the central battalion and of the one on each ſide of it will *file* and place themſelves in column, before and behind the ſtanding diviſion.—The other battalions will each be conducted in a ſeparate column to its point of entry on the new line, where it will throw itſelf into the general column, and wheel up into line when ordered.

———

When the prolongation of the new line interſects the body of the line and faces to either flank. { The line will break to the diviſion which ſtands in the point of interſection.—Every thing between that diviſion, and the flank which is to be fartheſt removed from the old line will make a change of poſition on the named diviſion,

and

and ſtand in open column on the new line; facing to the named diviſion.—All the diviſions that have ſo changed poſition will each countermarch by files: The line will then be prolonged, 'till the rear of the column arrives at its point.

Or, The part of the line which is firſt thrown into the new direction may ſo effect it by the Echellon march on the named fixed diviſion.—The whole will then wheel into open column and prolong the line 'till the rear arrives at its point.

Or, The named diviſion being placed with its pivot perpendicular to the new direction, and fronting the way the line is to extend, the reſt of the line breaks inwards and backwards towards it.—That diviſion is then put in march, and is itſelf followed in column by that part of the line whoſe flank will naturally firſt come to its ground: Fig. 107. The other part of the line moves on at the ſame time in a ſeparate column abreaſt of it, the whole being thus in a double column of diviſions as marched off from the center.—The head and the column immediately behind it, *halts* when its following flank arrives at its proper point, but the other column proceeds, and throws itſelf into open column in front of the named diviſion.—The line is formed by the wheeling up of diviſions.

S. 179. *Taking up Lines of March, and Formation.*

1. The general direction of any ſtraight alignement on which troops

are

are to form is always determined before they enter on it, and the point in that line at which their head is to arrive must next be ascertained.—Whenever the troops are to march on it, in column, or to form correctly, the line must be accurately traced out, and subdivided by mounted officers; and such officers when trained to that purpose, are the most general and surest points to move upon; particularly in situations where heights, and valleys intervene, and where no remarkable objects distant or intermediate occur in the direction, which (perhaps as relative to that of the enemy) must be chosen.—Another great advantage thence arising is, that although a distant object of march may not be seen or known but by those at the head of the column; yet the detached marking officers must be known by every pivot leader, to be there placed for the purpose of marching or forming upon.—Such officer, if he remains mounted, and which he always will do, when he can depend on the steadiness of his horse; will face to the line and have his horses head directly over it: If he is dismounted he will himself stand on and faced to the line; and with the hand which is farthest from the column, he will hold his horse by the head, and rather behind himself.

Lines of march and formation, best given by mounted officers.

---

2. Before a column of march or manœuvre approaches the ground, where it is to form, the commander will ascertain as circumstances may determine him, the advanced and distant points at which the flanks of his line are to be placed, or which he intends to be in the prolongation of the line when formed.—If he enters his alignement at one of those determined points, he continues his march straight upon the posted intermediate officer and the other point.—But if he enters the alignement, somewhere between them, it then becomes necessary

Distant objects of march, or formation.

fary to afcertain the fpot where the direction of his march interfects the new alignement, for at that point the head of the column arrives in it.

3. When the head of the advancing column approaches whatever part of the ground it ought to arrive upon.—Two officers, R, S, are fhown the flank diftant points of the alignement, P, T, and are fent forward to determine the intermediate point S, at which the head of the column ought exactly to enter into the new direction.—They feparate from each other 80 or 100 paces, go to the fide to which the column is not to wheel, and R, immediately places himfelf in the line of S, P, advanced before the head of the column.—They then both move on, R, always preferving, S, in a line with P, and each defcribing the portion of a circle upon P, as a center.—S, looks to R, and moves on, while the point T, continues to be advanced before him; but the inftant he has brought, R, in a line with, T, they both halt, and the 4 points are then in the fame line: R, remains fixed, till S, has fhifted to the point S, of interfection, and to enter at which the head of the column is now approaching.—This done R, alfo moves if neceffary to within 50 or 60 paces of S, and S, R, thus become a general bafe, which the appointed officers and adjutants immediately prolong for the march of the column, and in which they are affifted and corrected by the known diftant points.

Method of finding an intermediate point between two diftant objects.

Fig. 99.

This method of finding an intermediate point between two given, and perhaps inacceffible objects, muft be thoroughly underftood; and more than one column may in this manner afcertain their relative points of entry in the fame line.—Officers employed to give the direction

rection may with moderate practice take it up at the gallop, and therefore no halt, or stop of the column is to be apprehended.

To determine which of two columns arrive first at a given point.

Fig. 99.

4 When two bodies are in march to gain the same given point; the above method may most usefully be applied to ascertain which of them can first arrive at it.—The column B, and enemy D, are both in march on the point S. The leader of B. observes a distant point at C, beyond and a-head of the enemy D. If he can continue to keep this object open, and in front of the enemy, it is a certain sign that he approaches fastest to his wished for point; but if it appears as if moving towards the rear of the enemies march, it indicates his advantage, and the attempt must be given up in time.

S. 180. *When a considerable open Column — Enters — Marches — and Forms — on a straight Alignement.*

Necessary points of entry and march.

Fig. 101.

1. Before the head of a considerable column of march enters a straight line which it is to prolong and form upon; the point, s, of entry must be marked by a fixed person, who is to remain there till he is relieved; also another point, r, at least 60 or 70 paces from the first, and in the exact direction which is to be given to the new line. —Three other persons m, n, o, immediately and successively prolong themselves on r, s, as the original base, and being also corrected from s, upon such distant point (if any) as the commander shall have taken, they place themselves at least 200 paces from s, and from each other.

2. The

2. The line being thus in time prepared, the head of the first battalion arrives and wheels into the direction at s, and the adjutant of that battalion remains at the point of entry till the last division of his battalion has entered; he then gallops on for about 200 paces and posts himself on the line.——The adjutant of the 2d battalion who has in his turn placed himself at the point of entry, as soon as his last division has entered at it, gallops on and relieves the first adjutant, who goes on about 200 paces farther and again posts himself.——When the 3d battalion has entered, its adjutant relieves the second, that second relieves the first, and the first proceeds 200 paces farther and again alignes himself.—In this manner and till the last battalion has entered the line, do the adjutants successively and diligently relieve each other. If any of the adjutants overtake or interfere with the advanced persons who in the front are prolonging the line, such adjutants may return to their battalions, as being no longer of service.

Prolongation of the line by adjutants.

Fig. 103.

3. As to the persons m, n, o, who are in the front of the column: two of them at least having taken their station by the time that the leading battalion enters the line; as soon as the head of it approaches the first of them, he gallops on and anew alignes himself beyond the other two, and this operation each successively repeats till the column halts, which has always had (independant of any accidental distant point) two such persons to march upon.

Prolongers of the march.

Fig 103.

4. So many fixed points being thus ascertained, all which are successively passed by the pivots of the column, the accuracy of direction cannot but be preserved; but as a farther aid, each commanding officer of a battalion is at the point of entry invariably to place himself on the flank of his leading division, and in this situation moving on the

Commanding officers of battalions.

c
posted

posted adjutants, he (allowing for the breadth of his own horse (cannot fail to keep his battalion in the true line, by frequently going before his division, turning round, and correcting his flanks if necessary on the nearest adjutant in the rear, which the wave of a hand will suffice to do.—As each commanding officer arrives at a posted adjutant, he must go behind him, and again take up the flank of the division.

<small>Attentions in the march.</small>

5. Every division of the line having carefully taken its just wheeling distance before arriving, or at latest when it does arrive at the point of entry, and from thence having invariably preserved the step by a steady march; the pivots also (occasionally corrected backwards) having preserved the just line of the several adjutants; and no halt, or alteration of step, or distance, having been made by any one division or battalion from the instant that it has entered the line: The whole HALT at the same moment, on that word being loudly and rapidly repeated by each commanding officer, who immediately examines and corrects his pivots, and the column is thus prepared for the next order of wheeling into line.

<small>Adjutants.</small>

6. If the column halts when the last battalion has entered, the adjutants remain fixed, 'till the line is corrected and formed: but if the column is still carried on, then the person posted at the point of entry when he sees the rear of the column approaching the last adjutant, quits that point, relieves that adjutant, (who proceeds, &c.) and repeats that operation, 'till the column halts, and forms in line.

<small>Correction of pivots.</small>

7. When the column halts in an alignement to form; the various marked points in it which then exist, give the greatest facility to commanding

manding officers inftantly to correct their pivots if neceffary, each (as has been mentioned) on the next pofted adjutant in his rear; and which will alfo generally be on the pivot of the front divifion of the fucceeding battalion, in the fame manner as companies drefs from the pivot of one to the pivot of the next; for fuch correction fhould be meerly internal, unlefs fome inexcufable miftake has deranged the whole and thrown the rear of the column out of its true direction.

8. If great accuracy is required in the movements of a fingle battalion column, it is evident how much more effential it becomes in a confiderable one where faults would operate in the proportion of its extent, if they are not immediately prevented by the facility with which mounted officers can line, and correct upon each other.

9. When a column halts to form; fuch perfons as are then marking that line are not to quit their pofts, till fo ordered or till the line is put in march.

10. In marching in an alignement; if the rear or front of a battalion has evidently deviated from the true line, the head of the fucceeding one is not to follow its bad example, but muft preferve the general given direction into which the other is immediately to return. —And no commanding officer of a battalion when marching in an alignement, is on any account to alter the rate of march, or partially to halt, and thereby to derange the whole column.

11. Although the pofting and fucceffive relieving of adjutants on the line will undoubtedly preferve the direction; yet troops that are fufficiently trained ought certainly to prolong and form juftly on any line, *All in preferving the alignement.*

line, by having; 2 given points of march always a-head of the column, one point of entry marked and remaining, the commanding officers of battalions moving correctly on the flank of their leading divisions, and the adjutants or other mounted officers only occasionally stopping in the true line till the battalion they belong to has passed.—This should suffice to correct any small inaccuracy of the pivots, and keep the whole in the general direction given by the officers advanced in front of the column.

Step.

12. As the justness of step determines the accurate movement in column, that taken by the first leader must be frequently referred to and examined by the plummet, and every battalion marching in column should in order to regulate its march, have in its front a non-commissioned officer, trained and steadied to the equality of step.

Fig. 102.

Distant objects of march advantageous.

13. Where circumstances determine the march of the column on a conspicuous distant object, T, it is an essential help, and must be immediately declared to the leaders of the column, and as soon as possible looked out for, and remarked by all mounted officers; and if such another object also happens to be in its prolongation to the rear, it will aid in the correction of the march and in the formation of the line: but it will oftener happen that no such objects can be taken, and that the alignement depends on the direction (determined by relative circumstances) that the commander at first gives to two posted persons, and which is afterwards prolonged by others.

Fig. 103.

14. When part of a column is in low ground, or crossing a valley, its march can be directed and assisted by the rear points, at times when the front points of march are not to be seen.

15. No

15. No circumstance whatever is to occasion an encrease of the proper distance betwixt battalions in column.—The battalion guns will therefore march a-breast and always well closed up to the rear division of the preceding battalion; or, according to circumstances, they will move on one of the flanks opposite to their proper intervals; and if ordered on the front or pivot flank, they can occasionally fire if so required.——Music, pioneers, &c. are never in the intervals betwixt battalions in line or column of manœuvre, but are on the flanks of the column, or in the rear of the line. *Movement of battalion guns and justness of intervals.*

16. The most considerable column ought to be able—to MARCH in the alignement with perfect exactness; to HALT; to WHEEL into line; to MARCH forward; to HALT; and to FIRE; without more than a momentary pause between each operation, and without any necessity of dressing, correcting distance, or any alteration whatever; and unless the battalions are equal to, and can be depended on for such operation, no critical or advantageous measure when close to the enemy, can be attempted. *Correctness of movements.*

17. It is only when the column of manœuvre is marching in a straight alignement, that the commanding officer is invariably attached to the head of his battalion; for in other situations of march he must by no means remain fixed at its head, but be moveable on its flank in order to watch over its general progress. *Post of commanding officers of battalions.*

---

18. *When a line already formed is to wheel into open column, and prolong its direction.*—Three persons m, n, o, take their stations in the front as points of march, and the adjutants place themselves each close to the *Prolongation of a line.*

pivot

Fig. 103.

pivot flank of his own second division.—The column is put in motion, the last adjutant, when the rear approaches the one next to himself relieves him, and he going on they successively relieve each other.

Fig 102.

Change of direction.

19. *When a line prolonging a straight direction changes into another straight direction,* the advanced persons m, n, o, will of course be first placed in that direction, the front adjutant will be at the point of change, 'till he is relieved, and the column will proceed as before.

General aids on marching in column.

20. *If the march of the column* (although in open ground) *is not meant to be critically straight;* then the placing of adjutants can be dispensed with, and the divisions at their true wheeling distances will scrupulously follow the line which the head of the column traces out: but the better to prevent any improper deviation of the rear, commanding officers or adjutants will frequently stop at true points of the march until the rear of their battalion has passed, and always at points where the head of the column makes any considerable change of direction.—If every division of a column does not accurately follow the path traced out by the leading one, opening or closing of distances must take place, running up, or stopping short will ensue, and the column will not be in a situation to form in line with precision.

When the open column changes its situation on any fixed point within itself.

21. *When the open column of manœuvre has prolonged a straight line, stands halted, and is directed to make a change of situation on any fixed point within itself.*—All the divisions before that point countermarch and stand faced to it; the battalion if single, or the central battalion of a line and the

one

one on each side of it will file by divisions into the new column; the others will march in column, and enter where their *Rears* are to be placed.——If the column is intended to proceed, the division facing the given one having taken single distance, and the others of that wing being arranged behind it, they will all countermarch, and the column may then move on.—If the column is meant after such change of situation not to proceed but immediately to form in line, then the division facing the given one having taken double wheeling distance, ^{Fig: 45, 46} the line will be formed by a wheel up to the pivot flank. In either case a previous caution will determine the position of the division facing the given one.

S. 181. *Formation in Line on detached Adjutants, from the Assembly or Mass of Battalions in Columns of Companies.*

If a column of several battalions has halted at half, quarter, or close distance, or that its battalions have assembled in contiguous columns with small intervals; and that they are to extend into a line which is ^{Fig: 106.} at some distance from their then situation, on their respective adjutants, and facing either to the front, or to the rear.

2. A battalion is named as the one to be formed upon, and which may be either a flank or central battalion of the new line; but should be that one, which being placed at the point of appui determines the position of the line, and therefore will commonly be a flank one.—The general column when arrived at and standing on the new line, should always front to the point of *appui* whether flank or central.—Each

General preliminary attentions.

adjutant

adjutant marks one certain flank of his battalion in the new line; and each in taking up his ground allows for the front and interval of his own battalion from the laſt placed adjutant before him.——Each adjutant always marks that flank of his battalion in the new line, at which its head is to enter, and at which its REAR diviſion in column is to reſt, and therefore it is that flank which is fartheſt from the point of *appui*; if his battalion is to march with its right in front, he marks its left, and if with the left in front, he marks its right. It is therefore often neceſſary as will be mentioned, that ſome or all of the battalion columns ſhould ſeparately countermarch (S. 101.) at their point of aſſembly, in order to move off with their proper flank diviſions in front, and thereby enter the new line at their reſpective adjutants.

    3. Suppoſing therefore that the battalions are ſtanding in columns (the right in front) either in general column, or in contiguous line.—
Fig. 106.
If the poſition is to be taken from the *right* B, of the new line, the adjutants will from thence prolong it, each ſucceſſively marking his own left.——If to be taken from the *left* C,; the adjutants will from thence prolong it, each ſucceſſively marking his own right; and the
Point of appui determined.
battalions on ſeparating from the general maſs, will each countermarch, ſo as to arrive at its adjutant a column with the left in front.——If to be taken from a *central* point D,; both flanks of that battalion muſt be marked; its adjutant and thoſe of the battalions ſtanding to its left (or behind it if in column) will mark each his own left; the adjutants of the battalions to its right (or before it if in column) will mark each his own right, and thoſe battalions will in conſequence countermarch ſo as to enter with their left in front; and in this manner will the whole ſtand on the new line facing to the central point.

                                                                    4. Theſe

4. These circumstances determined and understood; all the adjutants are sent forward to the ground of the named battalion: the general direction of the line is ascertained by stationed objects: the flank point of entry is taken by the named adjutant: and all the others from him successively prolonging the line, mark their respective ordered points of entry; they are expected to give ground quickly according to circumstances, both by their eye, and their own step, as well as by the step of their horses. *Adjutants mark flanks.*

5. In the mean time, the whole are put in motion, and when sufficiently advanced, they HALT.—Such battalions as are to countermarch are ordered so to do, and each then diverges to right, or left, avoids crossing or interference, and marches quick to its own point of entry, opening its divisions in the course of the march.—At that point a momentary halt is made; the head division wheels into the line, the others successively follow it at open distances, and in ordinary time, (S. 125. 115.) till the word HALT is given on the arrival of the rear division at that point.—The battalion thus standing in open column, and its pivots being corrected on the adjutant, is ready to wheel up into the line, which is in this manner separately entered by each battalion, whether it is to face to the front or to the rear of the march. *Battalions enter the new line.*

---

6. As the adjutant always marks the point where the REAR division of his battalion column is to be placed, so the point where the head one is to rest will be of course easily known (and may be also marked in due time by another detached person) being at the distance of a proper interval and the front of a division from the preceding adjutant.—

The several adjutants when placed, become so many points of march to the battalions that are prolonging the line.

General rule. 7. Altho' unnecessary ground may seem to be gone over by the head divisions of some of the battalions, when they enter at their rear point; yet the rule that each of them shall enter the line where its REAR is to rest, is simple, general, and most readily corrects any mistakes that may be made; and all circumstances considered it is a quicker and surer manner of forming on the new position, than if the battalions were to enter at their head, or intermediate points.

8. When the enemy cannot possibly interrupt the movements of the detached battalions, this is an expeditious method of taking up ground in a defensive position: but it requires great exactness in the distances given by the adjutants, for if they misjudge their points there will be false intervals in the line, which can only be remedied by the battalions marching on to their proper distances before they HALT.—If the battalion of *appui* is nearest to the new line, and the first to form on it; then as all the others must enter it successively, any inaccurate marking of the adjutants may be remedied; because each battalion without interfering with any other one, can before it HALTS, march up to its just distance from its preceding one: but if the battalion of *appui* is the last to enter the line; each must then HALT at the point marked by its adjutant, and no correction can be attempted, till the battalion of appui has halted, and that the whole are in one general column.

9. Should adjutants be ordered to mark the head (instead of the rear) point of their battalion columns; and should such columns not countermarch, as is before required: In such case, each must sometimes

take

take diſtance not for the front of his own but for the front and interval of an adjoining battalion, and the column would not face to the point of appui; theſe circumſtances would much tend to embarraſs the formation of the line.—When an adjutant has to allow for the front of another battalion, he muſt be apprized of the number of files, officers included, in ſuch battalion.

10. Although the adjutant does always mark the REAR flank of the battalion column; yet as its head point, or any intermediate one is afterwards eaſily aſcertained, it can be directed (when particularly ſo ordered) to enter at either of thoſe points, as well as at the rear one, for any of thoſe operations places the whole in open column in the new line.

11. When battalions aſſemble in line of contiguous cloſe columns, they ſhould be ſo placed that no croſſing or retardment of the aftermarch may be occaſioned.—If the new poſition to be taken faces the ſame way as the columns do, the battalions ſhould ſtand in their natural order from right to left.—If the poſition to be taken faces to the rear of the columns, the battalions ſhould aſſemble in the reverſe order ſo that the right one ſhall be on the left; or if they otherwiſe aſſemble at firſt, they muſt countermarch in maſs, in order to ſtand ſo.

---

12. If poſitions are to be taken up to the front E, or to the flanks B, C.—The circumſtances already mentioned, will determine from what point the general line will be given, and what flank of his battalion each adjutant ſhall mark.—The ſeveral battalion columns (having countermarched if neceſſary, and if ſtanding in general column hav-

Poſitions taken to front or flanks.

ing difengaged into an Echellon pofition) march towards their adjutants, taking care to diverge to that hand which does not crofs the
path of the leading battalion or of each other; and when they approach the new line whatever way it fronts, each is in a fituation to
enter it at its REAR point, or, if particularly fo ordered at any other
given point.

Fig. 105.

13. If pofitions D, are to be taken in the rear.—Each battalion
will countermarch its divifions by files, fo that the columns ftand
with their left in front; the battalions then having the new pofitions
before them, will proceed accordingly.

Pofitions tak-
en to the rear.

14. As in changes of pofition, the arrival and formation of battalions in line is generally fucceffive; the head point of each can be readily afcertained from the fituation of the preceding battalion, even
before the whole of it may be fteadied in the alignement, and an under officer may in time be fent forward the more exactly to determine
it.—But the *rear* point of each, at which the adjutant places himfelf,
muft as to diftance often depend on his eye alone, and being mounted
he will always have fufficient time to take it up; as to the direction of
the line which is the great object of his attention, he can never fail in
it, if he takes it carefully from the prolongation of fuch objects as he
fees are placed in it, and of fuch part of the troops as may be formed
on the line.

15. The quicknefs and accuracy of all formations of the line, and
of all changes from one pofition to another, depends totally; on the
intelligence of each commanding officer who always conducts the leading divifion of his battalion to its point of entry in the new line; and
alfo

alfo of the adjutant who prolonging that line, marks the point of his laft divifion, and is himfelf the object on which the pivots of the column, or the divifions of the Echellon or column that fucceffively come into line are dreffed upon.—When the adjutant marks the *rear* point for the entry of the battalion column, he muft be accurate both in his diftances and direction: When he marks it only as a point of dreffing for divifions that fucceffively arrive in line, the juftnefs of direction is then the material object.

S. 182. *When the rear Battalions of a Column break from it, in order to enter, and form on an Alignement, in which the head ones have halted.*

If a confiderable open column has at any time partly wheeled into and prolonged a new direction, and that the head being arrived at its point, the whole are ordered to HALT with an intention of forming line in the new direction.——On the ceffation of march the entire battalion neareft the line, and any partial divifions of the one preceding it, that have not entered when the whole halt, fhall immediately by FACING and FILING, gain the new line.—But all the other battalions in the rear, fhall break from the general column, and each MARCH quick and feparate in individual column, till it arrives at its adjutant, who having expeditioufly lined himfelf on the head objects of the new line will be placed at its *rear* point of entry; the battalions will then prolong the line, and as they muft have fucceffively arrived in it, each will halt when its head is at a due diftance from the preceding battalion, its pivots will be corrected on its adjutant, and it will thus be ready to

Fig. 10.

wheel

wheel up into line, when the next battalion behind it shall have three divisions at least correctly standing in column on the line.——Or, the adjutants still marking the rear points to their battalions; if so ordered the head of each may be conducted to its respective head point (which is readily ascertained) it will then HALT, FACE, and FILE into the new line, and its pivots being corrected on its adjutant, it will be ready to WHEEL up into line as in S. 124.

A column marching at half, or quarter distance may in the same manner take up its ground.—The division that is to stop at the point of *entry* being ascertained, such part of the column as is before that division will successively there enter the line and prolong it at open distances.— In the mean time such battalions as are behind that division breaking from the general column, will march to their respective points and extend along the line.

S. 183. *When a Line of several Battalions, thrown into open Column, changes Position on a fixed flank Division.*

Fig. 47.

The direction of the new line, being ascertained, and prolonged, and the flank company placed perpendicular to it as already directed, (S. 120.) the whole wheel backward into open column, facing to the standing company.—The flank battalion FACES and FILES into column on the new line, (S. 120.) but the head division of each other battalion wheels and MARCHES off quick in separate column, to its adjutant who marks its *rear* point in the new line; it there enters, prolongs, and wheels up, each successively: as directed (S. 125.)—Or, if so ordered, each battalion may enter at its head point, as in (S. 124.)

S. 184.

S. 184. *When a Line of several Battalions, thrown into open Column, changes Position on a fixed central Division of any one Battalion.*

The direction of the new line being ascertained and prolonged, and the central company placed perpendicular to it as directed, (S. 122.) the whole line breaks backward into open column, so as to stand faced to the central company.—The companies of the central battalion and of the one on each side of it proceed to FACE, FILE, and place their pivot flanks in column on the new line (S. 122.)—But the head division of each other battalion, wheels and MARCHES quickly in separate column to which ever hand necessarily conducts it towards its proper *rear* point in the new line, which is marked by its adjutant, it there enters, prolongs, &c as in the preceding section.

Fig. 48.

In central changes of a battalion or line.—The movements of the right wing whether thrown forward or backward, are those of a column with the left in front, the rights being the pivot flanks; and the movements of the left wing, are those of a column with the right in front, the lefts being the pivot flanks.—In changes of position on the right of a battalion or line; the movements are those of a column with the right in front.—In changes of position on the left of a battalion or line; the movements are those of a column, with the left in front.

S. 185. *When a Line of several Battalions, thrown into open*

open Column, changes Position on a moving central Division.

Fig. 107.

1. The direction of the new line being ascertained and prolonged: the named company, a, will be wheeled and placed with its pivot flank perpendicular to and on the new direction, fronting the way the line is to extend, and if to the rear it must therefore countermarch.—The line will then break backwards by companies, so as to stand faced to the named company.—That company, a, will now be put in march along the new direction and be followed in double column by the remaining companies of the central battalion, and covered by one of those columns, viz. by that whose flank in prolonging the new line, will naturally first arrive at its proper ground, and which march with their pivots upon that line.—When the named division arrives at its new point, a, 2. it will together with those that are marching behind it, receive the word to HALT: such divisions of its battalion, which are to be in front of it, and are now marching by its side in column, and are separated from it by a distance of 3 or 4 paces, will move on, and by *filing* from their pivot flanks, will successively place themselves in column before and facing to it, at a double wheeling distance.

2. The other battalions —— which moved when the central one did; which in the mean time have been marching, each in separate column led by its inward flank division, and which have been pointing to front or rear, relatively to the movements of the central battalion, approaching, but not entering into its direction, except such as would naturally follow on the prolongation of the line: Those battalions will when the central one HALTS, march quick towards their several adjutants who

have

have been detached to mark their *rear* points, enter, prolong, and wheel up into line, as already directed.——In this movement some of the battalions near the central one might form to advantage on their head points, by filing from their pivot flanks into line, and if so ordered they may do it accordingly.

———

3. If the named company is a flank one of a central battalion: in that case the whole of that battalion will follow it in one column only, and the adjoining battalion will compose another column, and march a-breast of it, separated by 3 or 4 paces, 'till the named battalion comes to its ground and halts; the adjoining battalion will then proceed and by filing round from its pivot flanks, will (standing faced to the directing one) take its place in the general column, in order to form into line.

4. On many occasions where the named company is to be moveable, and that it is a central one of a battalion; the whole of that battalion if it is thought adviseable, may without much loss of time be thrown into one column before and behind that company, and the companies of that battalion which are in front of the named one will be countermarched, in order that the whole may face the way the column is to move.—This done, the general movement of the central and other battalions, each in separate column may begin; and in such case, the front company of the central battalion will be the first to arrive and HALT at the point where the column is to wheel up into line.

5. This movement of the given division is equivalent: to the line marching from the center either to front or rear, and from that situa-

tion forming away to the flanks: or, to the whole line first marching forward, or backward, and then making a central charge on a fixed point.—At the same time that it changes the front of a line, it carries the flanks to whatever point in that line it is meant they should rest at: it is the movement, which a second line does make, in order to comply with a change of position made by the first line, on a fixed point.

Fig. 109.

S. 186. *When the Head of a considerable open Column in March arrives at, or near the Point from which it is to take an oblique Position* (B) *facing to its then Rear, and at which Point its 3d, 4th, or any other named Battalion is to be placed.*

Fig. 109. B.

1. In general, the column after entering the new line would continue its march in that direction, till the named division arrived and was halted at the point of intersection; the battalions that had not entered into the line would then break off from the old direction and gain the new one.——But if such a column was marching parallel to an enemy's position, that its head had passed the enemy's flank point as far as was intended, and that the object was to take an oblique line and attack that flank: In such situation it might be too hazardous to allow the rear of the column which was destined to become the refused flank of the new line, to remain so long in its parallel direction, and it might be essential to draw it farther from the enemy as soon as possible.

2. Suppose the column consists of six battalions, and that it is determined

termined that the head of the 5th shall be placed at the point of interfection (d).—The column moves on and when the head of it arrives at the point (d) in the new line, the two, or any proportion of the leading battalions, may by the succeffive wheeling of their divifions enter it and march along it in the ordinary manner; but as foon as the leading divifion of the column does enter it, the 3d, 4th, and every other battalion breaks off feparately to the rear and march quick in columns to gain the new line: the 3d and 4th battalions affemble in clofe column a little beyond the point of interfection (d) and the new line; the 5th entering at its adjutant who marks its rear, forms in open column on the new line, with its head at the point (d), and all the other rear battalions form alfo relatively in open column on the new line.—The two leading battalions having in the mean time prolonged the line, when it comes to the turn of the 3d, it gradually takes its diftances, follows in open column, as alfo all the others, till the whole are ordered to halt, and the line to be formed by wheeling up.

---

3. The juftnefs of this movement depends; on the points in the new direction being taken up quickly and with precifion, on the previous determination that a certain battalion or divifion of a battalion, fhall pafs or halt at the point of interfection; and that every part of the column which is behind that battalion fhall throw itfelf into open column on the new line behind the point of interfection, ready to prolong or to form the line whenever it comes to its turn.

4. This movement will often take place in the change of pofition of a fecond line, and is performed by all thofe that are behind the divi-

sion which is to stop at the point where the old and new lines intersect. —And at all times when the open column changes into a direction on which it is to form, and that the division which is to be placed at the point of entry can be determined, it much facilitates the operation to make every thing behind that division gain the new line as quickly as possible, without waiting till the head of the column halts.

Fig. 109. C.

5. Suppose the column marching on a line parallel to an enemy's front, to have entered opposite one flank, and to be marching towards the other as if meaning to form in parallel line, but that circumstances determine to form in oblique line C, and attack the flank it has passed. —The column will be halted when the rear has arrived at a determined point; the direction of the oblique line, C, will be given; each division of the column will countermarch; the battalion that is to rest at the point of intersection will be named; the whole will be put in motion.—Two or three of the leading battalions continuing their march will by the successive wheeling of their divisions prolong the new direction; such following ones as are to be before the point of intersection, a, will assemble close to it; such others as are to be behind it, will at once march off quickly and separately to their point of entry in the new line; and stand in open column upon it in proportion as the head advances the whole will extend along the line in open column, be halted, and formed by wheeling up.

6. A line formed parallel to an enemy, may change situation in the above manner, by wheeling into open column, marching on to the point of intersection, and then taking up the new oblique position.

7. If

7. If a column moving parallel to an enemy, should stop and take up a new position on any point then within itself, such formation would be a central one and made either on a fixed or moveable division.

Close Column, of the Line.

1. The great object of a considerable close column is; to form the line to the front in the quickest manner possible; to conceal numbers from the knowledge of the adversary; and to extend in whatever direction the circumstance of the moment may require; which 'till it is nearly accomplished cannot be obvious to an opposite enemy: It is a situation for the assembly, more than for the march of troops: It is not formed until the head of the troops is arrived in column of whole, half or quarter distance near the ground where they are to extend into line.—The formation from close column into line is an original one, generally protected by cannon and cavalry, made at such a distance as not to be interrupted by the attempts of an enemy, and avoiding the enfilade of artillery. Its positions cannot fail to be truly taken.

2. The close column should not exceed 6, or 7, battalions; where there are more troops, it is best to form more columns if it can be done; therefore the columns of march may often be subdivided when they come near the points of forming into line, be directed upon them, and then closed up.

[38]

3. In general the battalion close columns before they begin to deploy, should stand 2 companies in front, and 5 in depth: In this situation the right company has its officer and his serjeant on its right flank, and the left company has its officer on its left, and his serjeant on its right. (S. 147.)

Fig. 70 72.

4. From close column the whole or any part of the body may be ordered to extend into line to either hand, as circumstances may require.

5. When the close column is halted, each battalion of which it is composed is 3 paces from the one before it.

6. A close column must loosen its divisions before it can march in front, and its changes of direction must be made circling and on a moving point, to enable its rear gradually to comply: If too great intervals should be made in the column, they can best be closed by a halt of the head.

7. Battalions standing in mass should be 6 paces distant from each other before deploying into line.

8. A close column of 2 or even 3 battalions, may occasionally deploy in the same manner as a single battalion does, and on any division; but in proportion to the number of divisions does the difficulty of execution encrease, and at any rate the formation will probably be inaccurate and defective.—Therefore when several battalions are halted in a close column, they do first deploy in mass on any named one, and thus stand in contiguous line of battalion columns, with any ordered in-

terval

terval between each. If the columns are of companies, the intervals will be equal to the front of a company and a half, they will then form columns of 2 companies each in front: and the whole will then deploy into line on any named divifion of any battalion.

S. 187. *When a Column of March (by Companies) of feveral Battalions; forms clofe Column, and then extends into Line.* Fig 114.

When it is found proper to fhorten the column of march, the rear divifions are ordered to clofe up to a certain diftance.—The leading divifion of the column either halts, or fhortens its ftep, and the rear divifions clofe up to quarter diftance: an interval of a company is referved betwixt each battalion, and the divifions when clofed refume the ordinary march. } CLOSE TO QUARTER DISTANCE.

ORDINARY.

When arrived within about 200 yards of where the line is to be formed; the head is halted, and the rear divifions move on to clofe column. } FORM CLOSE COLUMN. HALT.

The third or any other battalion is then named, as the one which is to give the ground on the line, and points in it are already marked out by the advanced adjutants. } BATTALION COLUMNS WILL DEPLOY ON THE 3d BATTALION.

The 3d battalion ftands faft, each other one in mafs FACES to its proper hand. } OUTWARDS FACE.

Each

Q. MARCH.	Each marches quick to the flank without opening out.
MARCH. HALT, DRESS.	When the 3d battalion is uncovered, it marches forward to its place in line, and halts at its given points.
HALT, FRONT. DRESS. MARCH. HALT, DRESS.	The battalions that are marching to the flanks as soon as they have acquired an interval of one company and a half from each other, will succeſſively HALT, FRONT DRESS—MARCH—HALT, DRESS, with the 3d battalion which is now on the line. Muſic, drummers, &c. are in the rear of each battalion column; and alſo artillery unleſs otherwiſe placed.
FORM GRAND DIVISIONS, &c.	The battalions being thus placed on the line with the above intervals, and in columns of companies are ordered to form columns of two companies in front as in (S. 147.)
THE LINE WILL BE FORMED ON THE 3d DIVISION OF THE 3d BATTALION.	The line being now prolonged to both flanks—A CAUTION is given that the whole will deploy on any named diviſion of any one battalion: for example on the 3d diviſion of the 3d battalion.
OUTWARDS FACE.	The two right battalions and the front diviſions of the third, FACE to the right, and all the reſt to the left.
Q. MARCH.	The whole MARCH quick to the flanks, except the named diviſion, which advances into the alignement, and the reſt of the 3d battalion proceeds to make a central formation on it. (S. 150.)

The

The other battalions continue their march 'till each arrives at the point where its inward flank is to be placed; and when each does so, such flank divisions, whether it is the front or rear one, HALTS, FRONTS, and occupies its place in line, while the other divisions proceed, and make their deployment upon it.—In this manner the battalions succesfively deploy (Sec. 148. 149.) obferving the general attentions already given. } —DIVISION HALT, FRONT, &c.

---

2. The points of marching and forming upon muft be well defined: The head divisions of battalions that move along the line muft do it accurately, and by no means get before it: The files muft march correct, and the beginning of the deployment of each battalion muft be well timed, otherwife the general line will be ill taken up.——The general line is that on which the battalion ftood before the deployment begun, and the feveral adjutants will carefully and quickly prolong it, each giving a point near to where the outward flank of his own battalion will extend.

3. The battalion columns FACE, and feparate from the general column by a command given for the whole by the chief; but each column is halted, fronted, and brought up into line by its refpective commanding officer —In like manner when the feveral columns are on their line of formation, they will FACE and MARCH by word of command from the chief; but each will be ordered to DEPLOY at the proper place by its own commanding officer.

4. When feveral battalion clofe columns ftand arranged along fide of
each

each other and are in concert to deploy into line.—The named one of formation only can be required to form, either on the front, a CENTRAL, or the rear division; but each of the others necessarily form, either on its front, or on its rear one, as the circumstances of situation demand.

5. After the column of march has closed up to quarter distance, the leading battalion may, when thought proper, be at once directed to its point of halting; and the others may successively diverge from the column, arrange themselves as before along-side of it, and double up to columns of 2 companies.—The line of battalions in mass being thus formed, at such a distance from the position it is to extend on as circumstances point out, may from thence advance on a front 1-5th of its extended one (as the several battalion columns are now 5 divisions in depth,) and may then deploy into line, as near to the enemy as appears safe.—In such state of deployment the troops have not much to apprehend, as they are in a situation to resist any sudden attack: nor until they do begin to deploy, can the enemy provide against, or determine what position they will take up; as 4-5th of their number may be thrown to either hand, and as an oblique direction may readily be given by the previous placing of the several battalion columns in such intended direction, which is an easy operation.

Fig. 106. E.

S. 188. *Oblique Deployments.*

Oblique deployment.

The deployment of the close column into a line OBLIQUE to the one on which its head then stands, may in some situations be required where circumstances do not permit of the previous operation of placing

ing the column perpendicular to such line. As when a wing is to be lengthened out but refused; or an enemy's flank to be gained by throwing forward one or more battalions which have advanced in close column behind the point of a wing; or when the nature of the ground on which the column stands demands a deployment that will give a support to a flank, or preserve the advantage of a position. *Fig. 112.*

Such deployments must be made, by the troops as standing in one column, and by the whole as if one battalion, according to the mode prescribed for it; they do not apply to battalions separated and standing in mass on the same line. If more than one or two battalions take up an oblique line, it will require great attention in the commanders to preserve order, and to form with justness.—Such formations are required on the front division of the column; hardly on the rear or on a central one, the attendant difficulties are sufficiently obvious: The column must be well closed up, and two companies in front.——Whenever circumstances permit the column to be placed perpendicular to its line of formation it must always be done; oblique formations are unavoidable exceptions.

S. 189. *If a Battalion close Column of Companies should be required to form the Square.*

The column being halted with the usual interval of one pace between the companies, receives the cautionary command to FORM THE SQUARE; on which the front half of the companies in the column, take one pace forward; the first company then falls back to the second, one pace; and the 2 last companies close up 1, and 2 paces to the com- *Fig. 103. B.*

pany

pany before them.——The whole companies make an interval of 2 paces in their centre by their sub-divisions taking each one pace to the flanks; 2 officers with their serjeants place themselves on each of the front and rear intervals; 2 officers with their serjeants also take post in each of the encreased intervals in the center of the sides; and a serjeant takes the place of each flank front rank man of the 1st division, and of each flank rear rank man of the last division; all the other officers, serjeants, displaced men, drummers, &c. &c. assemble behind the center of the companies which are to form the flank faces.——On the word OUTWARDS FACE; the 2 rear companies face outwards, and 4 files (supposing the companies of 12 file each) on each flank of all the companies (except the first and last) also face outwards, the whole lining with the flanks of the front companies, and dressing in ranks from front to rear.—At the word Q. MARCH; the 5th file from each flank of all the companies except the 2 first, and 2 last, followed by the front rank man of the 6th file, move up to right and left and respectively fill up the intervals between the flanks of their own and the preceding division; the remainder of the men of the side divisions arrange themselves to their right and left, forming close in the rear of their own divisions respectively——The whole thus stand faced outwards, and formed at least 4 deep, with 2 officers and their serjeants in the middle of each face to command; all the other officers, as well as serjeants, &c. &c: are in the void space in the center behind their companies; and the files of the officers in the faces may be compleated by serjeants, &c. &c. from the interior, in such manner as the commandant may direct.——The mounted field officers must pass into the center of the column, by the rear face if necessary, opening from its center 2 paces, and again closing in.——When ordered, the 2 first

ranks

ranks all round the column will kneel and flope their bayonets; the 2 next ranks will fire standing; and all the others will remain in reserve; the file coverers behind each officer of the sides will give back, and enable him to stand in the 3d rank.——Whatever is the strength of the companies which compose the flank sides, the whole of them will face outwards except their 4 center files, which are always reserved for filling up the intervals.

To reduce the square.—On the word FORM CLOSE COLUMN; the files that faced outward will come to their proper front, and the files that moved into the intervals will face about.——At the word Q. MARCH; the grenadiers take one pace forward; and the 2 rear companies take one and two paces forward, and then face about; the files from the intervals take their proper places; officers, serjeants, &c. will quit the interior, move to their several stations, and the companies that composed the flank faces will be compleated; the companies will also close inwards by sub-divisions one pace.

S. 190. *When several close Columns are formed from parts of the same Line,*

The parts of the line which are to compose each column are named; each battalion forms a close column on one of its own named divisions: The several battalion close columns march by a flank, and place themselves before or behind the directing battalion of that general column to which they are to belong.

S. 191.

§. 191. *If several considerable close Columns are halted at accidental Distances, but with their Heads dressed, and are ordered to form in one Line.*

Fig. 113. 114.

At whatever distance the heads of the close columns are halted from each other; the separate battalions will move up into line, each column upon its own named battalion: The point and division on which the whole are to form will be named: The whole will extend from it. The distances and commencement of movement will be taken from the named point, so that the outward battalions may move successively as it becomes necessary to preserve their distances from the inward ones.—Or, the adjutants taking their points from the given one of *appui*; the columns will in the mean time be previously so placed as not to cross or interfere in the march; the battalions of each will then disengage; march; enter; and form on the line.

§. 192. *If two Columns halted at open, half, or quarter distance are to exchange places.*

Fig 115. E.

The divisions of each will face inwards and file; when they have nearly approached each other, one of them halts, the other continues in march, and passes thro' the intervals of the halted one.—Both columns then move on until they arrive, halt, and front on the ground which each other occupied, and which has been properly marked and preserved for them: during this flank march, the heads of the files are kept nearly dressed, and are regulated in each column by the 2 leading divisions.——This operation is necessary when a line is to be taken up and formed on facing the reverse way to what the columns

then

then do; and if such line is in the rear of the columns they also countermarch their divisions by files, in order to enter and prolong it. This mode of columns exchanging situation which is equivalent to the passage of lines, may be required on several occasions.

S. 193. *When two Columns are to form in Line in any given Position.*

Points are prepared.—The columns by marching, countermarching, exchange of situation by files, or by whatever other operation is necessary, are brought up with their heads to the given points in the new line: The columns close up; the battalions disengage, place themselves on the new line, the division or divisions of formation are named, and the whole relatively deploy into line.——Or, points being prepared by the several adjutants, the battalions will disengage at a due distance, march on their respective points and form in line. Fig. 113.

S. 194. *If there are two Columns, composed each of parts of two Lines, which are to form.*

The battalions of the second line will halt, at a proper distance from the first, and deploy or form in line in the same manner as the first one does.—Or, if the first line is to form facing to the rear, the second one will have to proceed and to pass it, in order to arrive at its relative situation. Fig. 114.

If two lines, march off to the front in 2, 3, or 4 columns, each composed of part of the two lines; advance at certain distances from each other, to where their heads enter on 2 given parallel lines; wheel their heads to a flank into and prolong those lines to any extent: Then as the columns of each line have of course joined each other, the whole will be moving in 2 columns of lines, ready to form by a wheel up to the flanks, when the object of the movement is accomplished, which probably may be that of outflanking, or turning the flank of an enemy.

If two lines marching in columns of lines to a flank; are unexpectedly obliged to make front to that flank: Then the new lines will be composed each of part of the old ones, by forming up to right, and left.

Echellon Movements of the Line.

Echellon movements of a great corps.

1. The ECHELLON movements of a great corps place it in an advantageous situation; to disconcert an enemy; to make a partial attack; or a gradual retreat.—Different previous manœuvres must always have diverted the attention of an enemy, and prevented him from being certain of where the attack is to be made.—It may be formed from the center or from either of the wings reinforced: If successful the divisions move up into line to improve the advantage: If repulsed they are in a good situation to protect the retreat.—In advancing, the several bodies move independent, act freely, and are ready to assist:

In

In retiring, they fall gradually back on each other, and thereby give mutual aid and support.

2. The Echellons of a line are according to its strength, of one, two, or three battalions each.—Tho' their flanks seem multiplied they are not exposed, as they cover each other, and if they are far asunder they may be protected by artillery and cavalry relatively posted. Strength of Echellons.

3. Echellons seen at a distance appear as if a full line: being short, and independent lines, they can the easier march obliquely to outwing an enemy, or to preserve the points of appui to a wing; and such movement may not be perceptible to an enemy. Oblique march.

4. The Echellon may be formed direct from line, on a flank or any central division, either marching or halted, to front or rear. Echellon formed on any division.

5. The whole or only part of the line may be thrown into Echellon, and that, either to the front or rear.—In the first case with a view to gain the flank of an enemy, or obtain a cross fire; in the second to refuse or cover one's own flank. Partial formations in Echellon.

6. When the Echellon is unconnected with a line, the advanced flank or division regulates all its movements; when attached to a line, it must depend on the motions of that line. Directing point.

7. The same general principles of movement and formation apply to all Echellons similarly formed however great or small they may be, and whether they are acting to the front or the rear.

<small>General directing points of great Echellons in movement.</small>

8. Echellons of half battalions or less, move forward by their directing flank, which is always the one advanced from, or wheeled to.—Echellons of battalions move by their advanced serjeants—Echellons of several battalions move in line each by its own center, and the whole by the battalion next the directing flank.

<small>Change of directing flanks, or divisions of the Echellon.</small>

9. By at any time halting the Echellon the leading division may be changed and instead of one flank, the other may be made the advanced one: or instead of an Echellon formed from a flank it may be converted into an Echellon formed from the center; this is effected upon any named division, by the relative and perpendicular movements of the others to front or rear.——In this operation when the Echellon is a direct one, the divisions of it will exactly pass each others flank: when it is an oblique one which has been formed by wheeling, a part of each in passing will necessarily be intersected by the one preceding it, and must therefore double in passing, and afterwards extend into its proper place.

S. 195. *When a considerable Line changes to an Oblique Position by the Echellon March of Companies.*

<small>Fig. 77.</small>

1. *If the new line intersects any part of the old line.*—The battalion so intersected will make its change of position on that fixed point flank or central (S. 159. 161.); and all the others will march in Echellon whether forward or backward to their respective points in the new line, before they successively begin to form in it. (S. 162.)

2. *If the new line intersects the prolongation of the old line.*—A point will

will be given in the new line where the leading flank is to be placed. —The leading division will be wheeled fo that it may move perpendicular on that point, and all the other divisions of the line will wheel up the fame number of paces: the whole will march up in Echellon regarding their leading flank as a moveable center, and as each battalion arrives at the new line it will halt, and form in it by a new interior arrangement. (S. 157.)

Fig. 79.

In thefe changes of pofition, the whole Echellons of a line are fituated, and may be confidered relatively the fame as the platoons or echellons of a battalion: the whole move together and connected at the ordinary ftep; each battalion arrives fucceffively at its point in the new line, and each as foon as it arrives begins its formation on it.——So that whether it is the battalion or a line which fo changes, the march is made with precifion, and each Echellon forms up in fucceffion.

S. 196. *When the Line marches obliquely outwards in Echellon of Companies, and changes Pofition inwards to move upon a Flank which it has gained.*

The line formed and halted marches to the flank in Echellon of companies (S. 155.) forms in line parallel to the one it quitted (S. 156.) and if it then inftantly makes an oblique change of pofition, (S. 159.) it will be placed in a fituation to march forward with the greateft advantage on the weak point of the enemy. ——Or according to the diftance from the point of attack; the line after refuming its parallel fituation may move forward a given fpace,

Fig 80.

g 2

then

then make its oblique change of situation, and again march on in the new direction it has acquired, on the enemy's flank.

S. 197. *When from Line parallel to an Enemy, considerable Echellons advance from a Flank to the Front.*

Fig. 118. 119. The divisions of the line and the distance of Echellons being announced; the flank Echellon moves on; when it has taken the given number of paces, the next one follows, and thus successively 'till the whole is in motion; the whole halt, when the leading Echellon halts.

Fig. 118. 1. Two under officers from each following Echellon will march in the line of each preceding one, so as to stop (when the preceding one does) in its just prolongation, and at the points at which the inward flank and center of the following one is to be halted when it is required to move up into line, and whose position in such line must be thereby easily and accurately determined ——If the Echellon is composed of more than one battalion the others when such detached under officers stop, will send forward to mark also their several centers in the prolonged line.

2. It depends alone on the conductor of the leading Echellon when it halts, to give it such a direction that its prolongation shall pass before the enemy's front; and if the others are to move up into line, and are then within reach of the enemy's fire, it is evident how much care each must take, not to throw forward its outward flank, and be thereby exposed to an enfilade.

3. Not-

3. Notwithstanding every measure taken to obtain exact parallel lines, the following Echellons must, and on the march will be guided by and conform to the leading one; their great object is to preserve in moving on, their parallel and relative situations, their ordered distances, and proper flank interval: In this they are to act in the same manner as when advancing in line, and having the leading Echellon to guide them, together with the assistance of the mounted officers who attend to their movements, and prevent their outward flanks from being thrown too forward; they will execute with justness, this important manœuvre.—The preservation of intervals is also as essential an attention, as in the attack in line.

4. *When large Echellons* having marched forward are to wheel up to their advanced flank, and form in line oblique to the one from which they departed. The outward flanks which are to be the standing ones, must be halted as soon as each touches the line on which the formation is to be made: and for this purpose a line must be ready marked by advanced officers (prolonged from the leading Echellon) on which such flank is to halt, and on no account to pass it.—Each Echellon forms in line by a change of position on that flank: but if there is not a previous arrangement of distances, and a degree of doubling of each in proportion to the intended obliquity of the line, there will be encreased intervals between the Echellons.

Fig. 119.

S. 198. *When a Line formed on, and beyond an Enemy's flank, moves to the Attack in great Echellons.*

The Echellon which is then placed perpendicular to the point of the
enemy's

[54]

Fig. 124.

enemy's flank will move on, the rest will successively follow it, from each hand, and at their prescribed distance; the Echellons on one flank will be refused, and on the other they will advance beyond the leading one, to envelope the enemy.

From whatever situation of Echellons a body is placed in; a CHANGE in those Echellons may be instantly produced, by altering the leading one, and all the others immediately taking new relative positions to conform to it.

S. 199. *When a Line formed in front of and obliquely to the Enemy, is to move forward from a flank to the Attack in great Echellons parallel to the Enemy.*

Fig. 120. 121.

The number of Echellons and strength of each being ascertained and announced, the Echellons will naturally be formed to and led by the advanced flank.—As the oblique Echellons of a battalion are formed by the wheels of each company; so the oblique Echellons of a line are formed by wheeling up the 8th file of each flank company of each Echellon, a given number of paces and then correctly dressing the company to it: the other companies of the battalion or body which compose the Echellon, wheel each their 8th file half that number of paces, dress up to it, and the whole march and successively line upon the prolongation of the given one, proceeding as in the Echellon change of position of one or more battalions on a fixed flank. (Sec. 159.) Great pains must be taken in the correct placing of the flank directing division of each Echellon.

1. It

1. It must be observed that when the *Echellons* have been formed from the *oblique* line, so as to stand parallel to the enemy's front, they will be doubled behind each other in proportion to the degree of wheel made, and that were they to move directly forward to form in line with the leading one, a portion of each would be thereby cut off, and the general extent of the line reduced.—To endeavour by obliquing in the course of marching to rectify this defect, would be very difficult, and is hardly to be attempted: It must be remedied either by an early attention to taking the necessary and greater intervals than usual before forming the oblique line: or, before the whole advances, making the Echellons take ground to the flanks, and place themselves in their proper relative situations, as they would be when formed from parallel line: This done the whole may move on, either from the advanced or retired flank, and when proper, march up into parallel line. Should this not be done, part of each Echellon would of course be excluded on forming the general line, and must remain behind it.— The line may also be formed (provided the front Echellon halts in a situation that will allow it); by each other one at that instant making such a change of direction backward on its regulating flank, as will allow it to march perpendicularly forward to its proper point in the new line, where by another change of direction forward it will take up the prolongation of the leading Echellon.

2. Whether the original line is formed *parallel* or *oblique* to the enemy's front, the Echellons before marching are always to be placed perpendicular to the line on which they are to move.——From this situation a diagonal march on the enemy's flank may be made, and in such case large Echellons must be broke into companies; but it is an

operation

operation difficult in the execution, that would require much circumfpection, and if attempted too near would be very dangerous, as the flank thrown up is much expofed to the enemy's enfilade.

Fig. 121.
3. This ATTACK can be at once formed from a column of march, or the open column of a line, which is prolonging a direction upon, and oblique to that of the enemy.—The column will halt, wheel into line, and without any fenfible paufe the leading flank of each Echellon will wheel up parallel to the enemy, the other divifions of the line will each wheel the half of that fpace, and move on into their feveral Echellons, the whole will then be ready to advance led by any named Echellon.

The ATTACKS of confiderable bodies are almoft always conducted on the principles of the ECHELLON; there are few fituations where the whole could act at the fame time, or where it would be prudent or eligible fo to do: they are therefore made by fractions of a line well fupported and reinforced.

S. 200, *With refpect to the Enemy, and the intended Movement; the E-CHELLON pofition may be taken from the* { *Parallel Oblique Column,* } *Pofition.*

1. *If from the line parallel to the enemy.*—It is previoufly divided into the feveral Echellons which are to compofe it; and the diftance at which they are to remain behind each other is announced.——The reinforced

forced flank or center which is to attack is then ordered to advance; each Echellon of 2 or more battalions moves on when the preceding one has gained the ordered distance of (perhaps 100) paces, and thus being regulated by the head, act according to the event of the attack.

---

2. *If from the line oblique to the enemy.*—This position having been taken from the columns of march, or in the course of advancing in line; and the divisions of the Echellons being ascertained; they are formed by wheeling up parallel to the enemy and to each other: The advanced or retired wing reinforced may then proceed to the attack, and supported by the others will act according to circumstances.—— Fig. 120. 121. One may attack upon any degree of obliquity, and by absolutely refusing one wing, place it in a situation the more readily to protect a retreat should it be necessary, and which will be greatly strengthened if a point of appui can be given to such refused wing.

---

3. *If from columns halted perpendicular or nearly so to the enemy.*—Their heads are halted at given relative points, and given distances; the attacking bodies form in one or more lines; the others extend to the flank in Echellon, being separated perpendicularly a space equal to the Fig. 122. 123. distances they halted at in their several columns: This space is augmented if necessary when the whole move on, and lines of two or more battalions each are thus formed.—From the Echellon position by flank marching the order of column may again be resumed.

4. *The advanced Echellon* being arrived at its object, the attack begins, and the others attend the event.—If it succeeds they move up into line

h

to

to perfect it.—If it fails, each falling back on each, is strengthened and supported every instant of the retreat; this will generally be done by the Echellons in the course of retiring, at the same time making a gradual wheel backwards on the posted flank of the corps, from which the fire of artillery will much check and enfilade an advancing enemy.

———

5. *The second line* when there is one, follows in every thing the Echellon movements of the first.—The battalions make the same degree of wheel, preserve the same relative position, and serve as a support to the first; the attack of the second line moves on therefore at the same time with that which it is to support.——The Echellons of one or more lines are generally retired from 100 to 150 paces, each behind the one preceding of its own line.——When necessary the Echellons of the first line may retire thro' those of the second and be relieved in the attack.

———

5. *Where a line is passing a defile to the front, and from or near its center.*—After passing it may first form at the head of the defile, in the Echellon position; the several divisions are then ready to move up into line, or by wheels towards the flanks to form in oblique lines, and protect those flanks.—It may also in the same manner pass a defile to the rear retiring from the flanks by Echellon, while the center protects the movement.

———

7. *When the line has to advance a considerable distance in front,* it may occasionally be done with much convenience in a degree of Echellon position,

position, by each battalion being retired 5 or 6 paces or more behind its preceding one.——The battalion of direction is the leading one, which must march with the greatest exactness, and when so ordered, the whole can in an instant move up into line.

8. *A line B, formed parallel or oblique* to the enemy E, threatens and commences an Echellon movement from its left; but on the arrival of the left at a favourable point of appui C, the whole halt, and an Echellon attack from the right (which has been strengthened) then begins; this attack D, having been supported as long as is proper and having failed, the whole fall back in Echellon F, on the left which remains posted.—From this situation an oblique line G. is taken to the left and from the left, by each adjutant marking his own right in the prolonged line: the battalions successively again retire, and then break into column the left in front, march behind each other, enter at their adjutants, and take up the new line.—This position G, may be quitted by throwing back the left of the line; retiring by alternate lines; or in any other manner as circumstances may require.

MARCH OF THE LINE IN FRONT.

1. The chief object of every other movement is the quick and just formation into line when necessary, and the consequent advance of that line in front towards the enemy.——If the correct march of a *single*

<small>General intentions.</small>

single battalion requires so much attention and precision, it is evident that these must be redoubled to procure the just movement of a line, which is the operation that immediately leads to the enemy, and is the most difficult, and material of all manœuvres.——To hurry and bring up troops to the attack in imperfect order is to lose every advantage which discipline proposes, and to present them to the enemy in that very state, to which after his best efforts he has hoped to reduce them.

2. The same principles that direct the march of the battalion direct that of the line; besides which several peculiar observances are required, and in proportion as difficulties encrease, must attention be given.

3. No body of troops can advance in line with firmness and order, unless the original formation of that line has been perfectly straight, and its correct preservation during the march requires every attention.

4. The cadence of the march is not to be altered by particular battalions; but when it is necessary each will lengthen or shorten its step by word from its own commander.

5. The march, and halt, and attention in line of the officers and men of each battalion, are by its own center; the commander alone regards the regulating battalion.—Dressing to a flank is by a separate direction, and given when necessary and proper after halting.

March.

6. Battalions in line, marching over heights, or across valleys, will require more time to pass them, than others who are moving on the same

same extent, but of level ground; in order to preserve equality of front, the last must therefore in general be ordered to shorten their step.

7. The march of a confiderable body in line can only be at the ordinary step, a quicker movement would produce disorder, nor could artillery well attend its motions when advancing to the enemy: But there are situations, where a brigade or smaller front should move on to a particular object or to an attack at a lengthy step, or where even a quicker cadence may be required from them. General pace.

8. When a line of several battalions is formed and halted; there is an interval of 12 paces between each for two pieces of artillery; the men are generally dressed to a flank; ranks are closed up; the whole stand ready for movement; and for which the directing serjeant of each has prepared himself as required in the single battalion. Intervals.

S. 201. *When the Line is to march in Front.*

One of the battalions is *named* as the regulating one, to whose movements all the rest are to conform.—The *Commander* of the line is himself with that battalion, every precaution as already prescribed is taken to ensure its perpendicular march, and its directing serjeants are ordered to advance.

At a *Caution* quickly circulated, that THE LINE WILL ADVANCE, the directing serjeants of each other battalion, move out their 6 paces; it is almost impossible that they should not halt in perfect line; but if any small alteration is necessary, THE LINE WILL ADVANCE.

ceffary, the ferjeants on either fide of the regulating one, being from that laft, ordered to move forward or backward as much as appears neceffary for this correction, will, together with the regulating one, give a line to which all the others will immediately conform.

The *Directors* of the march being thus placed parallel to the line, muft take care that their bodies are perfectly fquare to the front, and they will again remark their near points of march, for they muft not look out for diftant ones, but take fuch as accidentally occur on the ground; thofe of the battalion of direction are not liable to be altered, but all the others are to be confidered only as relative helps to begin the march, fubordinate to thofe of the regulating battalion, and liable to be changed from the inftant that they appear to the commander of the battalion to produce a movement which does not correfpond with the regulating one, whofe march is here fuppofed accurate, and as juftly taken as poffible; for, fhould that battalion take a falfe direction, univerfal diforder muft take place, unlefs it immediately affumes and perfeveres in a true one, perpendicular to the front of the general line.

From the circumftances in which lines muft generally be expected to move, as thick weather, fmoke of cannon, duft, &c. &c. it is evident that *diftant* objects of march cannot be looked for or taken, nor any other obferved than fuch as are near, and derived from the eye and the fquarenefs of the body moving upon them.

The *Commander* of each battalion is ten paces behind the rear rank, in the file of the directing ferjeant, and will there remain; his adjutant

tant is behind the flank next that of direction, and the major is behind the other flank.——The commander of the line, or some person that he appoints, is near the directing serjeant of the regulating battalion; and with coolness and judgment, may make such signal to the serjeant of a neighbouring battalion, as will gradually bring him forward or backward, and by then being a direction to the others, will tend to preserve the parallelism of the line, but this correction is not to be attempted, without great and gradual discretion, and so as not to occasion any considerable alteration to a flank of the line.

At the word MARCH given to the battalion of direction, and rapidly repeated, each battalion at the same instant is put in motion, by its respective center. } MARCH.

From the first moment of movement the quickest and greatest attention must be given by the commanding officer, to observe whether the direction of the regulating battalion is just; this will be seen in the course of 20 or 30 paces; for, if the rest of the line is moving steadily, and that this battalion is closing to one flank, and opening to the other, its direction must be changed accordingly by advancing a shoulder; but if the whole are steady, or that the battalion is not altering within itself, or with respect to the general line, its direction must be persevered in, and not afterwards changed. 'Till this circumstance is ascertained with respect to the directing one, and which must be immediately done, its contiguous battalions will make no alteration in their position.

The *Regulating* battalion must be regarded as infallible, the commander

mander of the line watches over it, and from the moment that its direction is ascertained, the commander of each other, and their directing serjeants, are to consider their movements as subordinate to it, and to conform accordingly: It is the helm which guides the line, and must not change cadence, nor will it lengthen or shorten its step, but from unavoidable necessity, and by particular order.

3. The instant communication of the word MARCH is particularly important, that the advanced serjeants of the whole may step off together, and thereby maintain their line parallel to the one they quitted, and which becomes the principal guide for their battalions; each preserves its 6 paces from its advanced serjeant; this distance is to be kept by, and depends on the replacing officer next to the colour, who covers the directing serjeant; and if these trained serjeants do step equally, and in parallel directions to each other, they must be dressed themselves in line, and of consequence the centers of their following battalions.

4. But as the *Flanks* of battalions are apt to be behind their centers, the majors and adjutants will particularly attend to this, and also the flank officers of each battalion, who being unconfined in their persons, may preserve themselves in the general line of the colours.—When a flank officer observes that the line drawn from himself through the colour of his battalion, passes before the general line of colours, he may conclude that he himself is too much retired; but when such line passes behind the line of colours, he may conclude he is too much advanced, and will regulate himself accordingly.—The great object in movement is to have the whole of each battalion perpendicular to the direction it marches upon, the whole of the several battalions in one

straight

straight line, and their several marching directions parallel to each other.

5. As the movement of the *directing* battalion is infallible, and must be conformed to, and as the preservation of intervals is the first and principal attention in the march in line, it is to that object, and of consequence to the direction of the march, that the exertion of the commanding officer of each battalion must be turned, and therefore the preservation of his interval from the directing hand is what will determine, and regulate every alteration he orders, and from the warning of his adjutant he will be always apprized when the interval begins to encrease or diminish. As to the other flank, he need never look towards it, that must necessarily follow and accommodate itself to the colours, under the correction of the major; but previous to making any alteration, he is quickly and decidedly to observe, whether the error arises in his own battalion, or whether it originates in one nearer than himself to that of direction, and which the battalion that alone has closed or opened to one flank, will naturally remedy by the counter movement, without affecting the order of the rest of the line.

6. The *regulating* battalion being supposed on the *right*, the commander of any other subordinate battalion, who finds himself closing the interval to the right, and that he ought to correct it, will instantly order the directing serjeant, RIGHT SHOULDER FORWARD; or if he is opening from the right, he will order, LEFT SHOULDER FORWARD; these changes the serjeant makes by a small but gradual alteration in his own position, and of course must change his points of march towards the ordered hand, the degree of such change it is impossible to ascertain by words; but by the subsequent movement of the battalion,

i

the

the commander muſt farther correct it, if neceſſary; it muſt in all ſituations be very ſmall, and will be proportionally greater or ſmaller, as made ſooner or later after beginning to advance, or from the laſt time of correction.——If for example, before he has marched 20 or 30 paces, the ſerjeant is ordered to change his direction he may conclude that he had taken one conſiderably wide of the true perpendicular; whereas if it is ordered after he has advanced 100, or 150 paces, he may judge that he has deviated but little from it. When ſuch change of direction has effected the firſt object, a very ſmall *counter change* will be generally required to preſerve the diſtance gained and the required front.——The replacing officer and colour in the center of the battalion, will on each command of change given to the directing ſerjeant, make ſuch relative movement as is neceſſary to correſpond with his new poſition.

7. When a battalion is marching in a true direction, but that occaſioned by the fault of others, an opening from, or cloſing towards the regulating battalion comes from that hand and muſt be complied with, the word OBLIQUE, (to right or left) is given; the battalion without loſing its parallel front, or eyes being altered, obliques, till it receives the word FORWARD, when the croſſing of the ſtep ceaſes; and the directing ſerjeant proceeds in full front, but in a line parallel to the one he quitted, and removed from it as far, as the line did incline.

8. Should a battalion from any partial reaſon be behind, or before the line.—— It will receive the words STEP OUT, or, STEP SHORT, and when the line is regained the word ORDINARY.

9. In *correcting* the movements of battalions in the line, much judgment

ment muſt be exercised, and wherever the fault does originate the remedy ſhould in general from thence begin.——The MARCH and HALT and attention of each battalion in line is by its own colours, the commander alone is obſervant of the regulating one, and it is only from the centrical ſituation preſcribed to him, that he can truly judge, and remedy the beginning of defects.——The major, and adjutant, by being cloſe to the rear rank, can keep up the flanks.

10. A battalion which is near to the point of appui, or the point of attack, will in general be the *regulating one*, therefore a flank battalion will commonly direct the movements of the line, and ſhould the commander change it, he muſt announce ſuch change.

11. In the courſe of marching ſhould an *obſtacle* **break** the center of the regulating battalion; immediately before ſuch interruption takes place, one of the battalions near it, muſt be named to the adjoining ones (but not neceſſarily to the reſt of the line) as replacing it, and may continue to direct in future, or, at leaſt till the colours of the former one have after paſſing the obſtacle, again regained their true poſition in the original direction, and which by the operation of detached perſons muſt have been truly found and traced for them.—— Whatever impediments preſent themſelves to the march of the line, will be avoided by the peculiar battalions, according to the modes already preſcribed, and the openings made by ſuch parts as are obliged to quit the line, will be carefully preſerved, in order to their re-entering into it, as ſoon as the ground permits.

HALT.
{ The *Line* thus marching in perfect order is at every moment prepared to receive the word HALT given to the regulating battalion, and in the most instantaneous manner circulated by the commanders of each other battalion, who constantly looking towards the regulating one, can lose no time in its repetition.——The whole halt firm at the instant the word is given, and no dressing or correction of intervals take place (till so directed), but the line should be ready immediately, and without farther preparation, to commence its FIRE.—The advanced serjeants on the halt fall back to the battalions.

S. 202. *When the Line is to dress.*

DRESS.
{ If the *Commander* gives the word DRESS, it is immediately to commence from the center of each battalion, the men looking to their own colours, and the correcting officers lining them upon the colours of their next adjoining battalion; the platoon officer on the left of the colours performs this operation for the left wing by placing his own platoon in the direction of the colour to his left; and the officer on the left of the right wing (or if there is none such the center directing serjeant) performs this for the right wing, by placing the platoon beyond him in a line with the next right colour; this done without delay, and without too much nicety, the wings of each battalion immediately conform to their two placed platoons, towards whom they are then looking.

By

By this means, when a single battalion halts; it is dressed on its right center company, and is therefore in a straight line.——Two battalions thus dressing from their several centers on each others colours, and their outward wings conforming, must therefore be in a straight line. ——When 3 or more battalions thus dress from the center of each on its next colour; if all the colours happen to have halted in a line, the general line will be straight; but if they have halted irregularly, then the portions of the line between each 2 colours will be straight and no flank will be exposed, which is giving it the best firing situation that dispatch and circumstances may allow.—In this operation the two center dressers of each battalion must be very alert.

But if the commander finds it necessary to give a more exact *dressing*,—he immediately orders the first colour of one of the adjoining battalions to move out two or three paces, to be planted upright, and the bearer to FACE towards him. ——He then himself advances a pace or two, and plants the colour of the regulating battalion, so that the line of the two prolonged, shall occasion as little change as possible at the distant flanks, consistent with his views; the colours and flank officers of the other battalions instantly move into that line, all the colours facing to the regulating battalion, and the flank officers to their own colours. So many fixed points being thus ascertained, the platoon officers are immediately ordered to cover in it, facing to their own colour, and the men of each battalion to move up as prescribed for the single battalion.—The advantage that arises, if platoon officers can at once take their covering and distances from the left, has been already mentioned.

DRESS.

The

The line may also be correctly dressed; by one colour of a battalion near the directing one advancing a few paces; one colour of the directing battalion is then placed in the intended line; a colour from each other battalion immediately prolongs this line.—The grand divisions of each battalion are then successively dressed from their own colour upon the adjoining one.

With practice and alertness, the *dressing* of a line of very considerable extent, may be quickly and readily accomplished; but the correction of an improper interval is not to be done without the side movement of every thing beyond it, which is no easy operation, and shows the necessity of the most indefatigable attention being given to this object; every encreased interval presents a weak point, which is studiously to be avoided; and every ill dressed line in movement will naturally create such intervals; nor is such a line in a proper situation to march up to an enemy whom its fire may have shaken, for disorder must attend its unconnected movements.

S. 203. *When the Line is to retire.*

When the Line is to retire, the necessity of its being previously correctly dressed, is full as essential as when it is to advance; if that preliminary is not taken care of, its movements must be disordered in proportion to its extent.——The several battalions will prepare for the retreat in the manner prescribed for the single one; by receiving the caution that the *line will retire;* and then by *facing* to the *right about*.

But as there may not always be time to give it the wished for

for degree of exactness, before the *Retreat* begins, such aids may be applied, as will greatly assist it, in the course of its movement.——On the caution that the LINE WILL RETIRE, the directing serjeants, &c. move to the rear six paces, taking their several directions as already prescribed; one of the colours next the regulating battalion will be considered as a fixed point, the colour of the regulating battalion will then be placed so that the prolongation of the line which unites the two, shall give the required front of the march, the other advanced serjeants and colours, without regard to distance from their respective battalions, will place themselves on this line.—The line then FACES to the RIGHT about.—At the word MARCH, the whole move on, the advanced serjeants preserving their position and line, and the battalions by degrees acquire their just distance of six paces from them. {THE LINE WILL RETIRE. R. ABOUT FACE. MARCH.}

When the *Line* in moving to the front, *halts* and renews its march, without any previous dressing, the same method may, without delay, be employed to regulate its advanced serjeants, before the word MARCH is given.

When the Line is to front, each battalion receives the word HALT—FRONT, and immediately faces about; if it is then to move forward, the colours and serjeants are ordered to advance before the front rank, and are there correctly lined, ready to conduct its march.—But if it is to remain halted, the dressing is then ordered, in the manner already prescribed. {HALT, FRONT.}

1. Not-

1. Notwithstanding every direction that can be given for the *march* of the *line* in *front*, the success of its execution will totally depend— on the complete dexterity, and training of its component parts; on the quick eye, and ready decision of the commanders of battalions; on the accurate cadence, length of step, and lines of march taken by the several advanced serjeants, and by the battalions; together with the perfect squareness of each individual's person; all these justly combined, are necessary to procure that precision which is not unattainable and is so essential in this most difficult and important movement.—— For the halt taking place near to the enemy, and when the firing should begin, there is no time then to rectify errors, and redress the line, but every thing must remain in the situation of that instant; and though a line a little irregularly halted, may not be deficient in fire, yet it will present exposed flanks of battalions, and will not be in a state to advance farther without disorder, or without first correcting its front, even should an enemy give way.

2. Although no *Caution* should ever be given before the HALT of the single battalion; yet it might sometimes help to assist that of the line, if there was not a danger that in permitting it, the march of the line would be habitually made with less accuracy, trusting to this correction, than if an instant *halt* was always to be expected; for this ought to be required of a line of any given extent; and experience has shown that it is to be attained when made with great attention, and on just principles.

3. *Echellons* of attack have generally a considerable front, as of three

three or four battalions, and equal to that, of the part of the body, against which they are directed.——But was a *Line*, in order to render its movement eafier, to advance up to an enemy's line in echellon of battalion, retired five or fix paces behind each other; although perhaps the intervals might be taken with fufficient correctnefs, yet in fuch fituation flanks would be liable to be thrown up, the general line not fo well attained, and the battalions as they halted would be expofed to an oblique, as well as direct fire, till the others came up.—— The firft part of a movement in front may be made in *Echellon* of battalions, but the *Line* fhould be completed before the laft fteps of it approach near to an enemy; for nothing can then fupply the place of, or be depended on, but the accurate *March* in *Line*, acquired from attentive habit, and juft training.

4. *A change* in the direction of a line, when the whole is in movement, can never be confiderable; and muft be made gradually, with great attention, and on the fame principles as thofe of the battalion. (S. 169.) Change of direction.

5. When the line is advancing in full front, or in Echellon for any confiderable diftance; the *Mufic* of one regulating battalion may at intervals be permitted to play for a few feconds at a time; and the drums of the other battalions may be allowed occafionally to roll. Mufic.

———————

6. Altho' a fingle battalion may, by opening its companies and files, *from* 3 *deep form* 2 *deep*, by introducing its rear rank into the other two.——Yet a confiderable line pofted, which is to be lengthened out to one or both flanks by its rear rank; muft to greater advantage Rear rank lengthening out a line.

k perform

perform such operation, by each company wheeling the sub-divisions of its rear rank backward, and facing to the hand they are to march to; the last rank of each company closes up to its first; the sub-divisions of each battalion move up to open distances from their respective head ones, and from each other; officers from the rear are appointed to command them; those of each, or of every two battalions being considered as a battalion, they march on in column and prolong the line. By this mode of lengthening out the line, the 2 front ranks remain undisturbed, and they protect the movement which is made unseen behind them.

S. 203. *When a considerable Line has to pass a Bridge or Defile.*

Line passing defiles.

1. It will proceed in the manner prescribed for (and as if it was) a single battalion, in passing to front or rear.

2. When there are several bridges or defiles to be at once passed; the line will be divided into relative portions, each composing a passing column.

3. When such movements are performed in presence of an enemy; the safety and protection of them by artillery and posted troops is a matter of previous disposition.

Advancing lines.

4. As the lines of infantry are generally formed not nearer than 12 or 1500 paces from an enemy's position, unless peculiar circumstances of ground favour a closer approach: There will often therefore be

be such obstacles to moving on, as besides the partial ones occurring to parts of battalions, may oblige large portions of the line to be again broken.—This becomes a matter of particular arrangement, in which the great object is the subsequent quick formation of the troops; and where considerable close columns or redoubled lines of battalions at small distances will be employed.—Such situations are always hazardous if the enemy is so posted as to profit of the movement of passing; or if favourable ground, and a superior artillery do not oblige him to keep at a considerable distance.

5. Where defiles are to be passed in approaching the position of an enemy; the mouths of them must be strongly occupied; the columns of march must pass on the greatest front they will allow of, be previously arranged, well closed up, and ready to deploy in an instant. *Columns of march.*

6. When in presence of an enemy, a retiring line is to break, and to pass defiles in one or more columns; much steadiness and disposition is required.——The line approaches near to the defiles before it fronts; the object is to pass quick, but without disorder; commanding grounds are previously occupied, and flanks are covered; if bridges are to be passed, protecting artillery will be placed on the other side. *Retiring lines.*

S. 204. *When the Line advances or retires by half Battalions, and fires.*

1. *If the line is in march and advancing.*—On the order from the officer commanding the line, the left wings, HALT, and the right ones continue to march 15 paces; at which instant the word MARCH being

given to the left wings, the right at the same time are ordered to HALT, FIRE, and load, during which the left marches on and past them, till the right wings being loaded and shouldered, receive the word MARCH, on which the left ones HALT, FIRE, &c. and thus they alternately proceed.

2. *If the line is in march and retiring.*—The right wings are ordered to HALT FRONT, and when the left ones have gained 15 paces, and receive the word HALT FRONT; the right wings are instantly ordered to FIRE, load, FACE ABOUT, and march 15 paces beyond the left ones, where they receive the word HALT FRONT; on which the left wings FIRE, &c. and thus alternately proceed.

General attentions.

3. In addition to the battalion directions (S. 177.)—There must be a regulating battalion named, by the half battalions of which, each line will move, halt, and fire: The commander of each line will be with such half battalion; and in giving his several commands must have an attention to the general readiness of the line, especially after loading, that the whole are prepared to step off together at the word MARCH. —The firing of the advanced wing succeeds the MARCH, or the HALT, FRONT of the retired wing instantly; and each half battalion fires independant and quick, so that no unnecessary pauses being made betwixt the firing words, the fire of the line should be that of a volley as much as possible; and the whole being thereby loaded together, will be ready for the next command of movement.—In these firings of the line advancing or retiring, the 2 first ranks will fire standing, and the rear rank support their arms.

4. In this manner also may the alternate battalions of a line advance or

or retire, and when the whole are to form, and that the laſt line moves up to the firſt, every previous help of advanced perſons will be given to enſure its correctneſs.

S. 205. *Firing in Line.*

1. The chief object of fire againſt cavalry is to keep them at a diſtance, and to deter them from the attack; as their movements are rapid, a reſerve is always kept up.—But when fire commences againſt infantry it cannot (conſiſtent with order, and other circumſtances) be too heavy or too quick while it laſts, and till the enemy is beaten or repulſed.

Object of fire

2. The fire of 3 ranks ſtanding is hardly with our preſent arms to be required; eſpecially if the ground ſhould be broken, and that the ſoldiers are loaded with their knapſacks.

3. Where infantry are poſted on heights that are to be defended by the fire of muſquetry; the front rank will kneel, that one third of the fire that may be given ſhould not be loſt, for otherwiſe the rear rank in ſuch ſituation could not ſufficiently incline their pieces to raſe the ſlope. —— As ſoldiers generally preſent too high, and as fire is of the greateſt conſequence to troops that are on the defenſive, and who are poſted if poſſible on commanding grounds, the habitual mode of firing ſhould therefore be rather at a low level than a high one; and the fire of the front rank kneeling, being the moſt efficacious as being the moſt raſing, ſhould not be diſpenſed with when it can be ſafely and uſefully employed.

Defenſive fire

4. When

4. When infantry marches in line to attack an enemy, and in advancing makes ufe of its fire; it is perhaps better to fire the 2 firft ranks only ftanding, referving the third, than to make the front rank kneel and to fire the whole: but volleys fired at a confiderable diftance, or on a retiring enemy, may be given by the three ranks, the front one kneeling.

In line advancing.

5. A line pofted, or arriving at a fixed fituation will fire by *platoons*, each battalion independant, and fuch firing generally commencing from the center of each.—The firft fire of each battalion will be regular, and eftablifh intervals; after the firft fire, each platoon fhall continue to fire as foon as it is loaded, independant and as quick as it can, 'till the battalion or line is ordered to ceafe.

Platoon firing.

6. Behind a parapet, hedge, or abbatis, the two firft ranks only can fire, and fuch fire may be *file firing*, deliberate and cool, the 2 men of the fame file always firing together: It may begin from the right or left of platoons, and fhould be taught in fituations adapted to it, not in open ground.——Should the parapet, hedge, or abbatis be but little raifed, platoon firing may be ufed.

File firing.

7. *Oblique firing* by battalions, is advantageous on many occafions; as when it is proper or that time does not allow to give an oblique direction to part of a line, or that their fire can in this manner be thrown againft the opening of a defile, the flanks of a column, or againft cavalry or infantry that direct their attack on fome particular battalion or portion of the line.

Oblique firing.

8. As long as the fire by battalion, half battalion, or companies, can

can be kept up regular, it is highly advantageous and can be at any time stopped; but should file firing be allowed and once begun, unless troops are exceeding cool and well disciplined, it will be difficult to make it finish, and to make them advance in order.

Regularity of firing.

9. When a line halts at its points of firing, no time is to be lost in scrupulous dressing, and the firing is instantly to commence.—But a line that halts and is not to fire, or when its firing ceases after the halt, may immediately be ordered to dress from colours to colours.

10. The attention of the officers and non-commissioned officers of the rear to the locking up of the ranks in firing cannot be too often repeated.

S. 206. *When the Square or Oblong is composed of more than one Battalion.*

1. Two battalions may form column of companies, each behind its inward flank one, and close the interval between them. The oblong when formed will be six deep, if the companies are at half distance and wheel outwards by sub-divisions, except the 2 first and 2 last which close and face outwards: The grenadier and light companies may be considered as making no part of the oblong, but be applied according to circumstances.

Fig. 110. D.

2. If a greater degree of space is required in the interior, of the figure; the side columns will be of sub-divisions, and formed behind the 2d or 3d company from the inward flank, a proportional number

of

of the last companies will compose the rear face, the oblong when formed will be 3 deep, and the grenadier and light companies will be in the interior, to be applied as is found necessary.

Fig. 108 C.
110. C.

3. *If 3 or more battalions are to form a square or oblong.*—It depends on circumstances what part of the line whether flank, or center composes the front face, but the line is subdivided accordingly; and the parts of it protecting each other in the movement, march in echellon or column to take up their different situations. The flank faces when the square marches will move in columns of subdivisions.——*When the square is halted;* the 2 flank subdivisions of each face wheel back the quarter circle and obtain a cross fire at the angles: The grenadier and light companies are ready to reinforce any particular part.——*When the retiring square* has arrived at a point where it can diminish in safety, the rear face may halt and front inwards; the side faces continue their march in column and their heads if necessary unite; the front face, halts fronts, when it arrives at the rear face; that face then retires in 2 columns, followed by the front face, which in this operation is protected by such posted companies as make the rear of the whole.—— *If the retiring square* arrives at a favourable position as a wood, heights, &c. which it is to occupy, the flank faces extend along it in columns of sub-divisions, protected by the other 2 faces, which afterwards take up their own ground. These operations being supposed of necessity, will invert parts of the line, which must be afterwards remedied as opportunity offers.

4. *The square or oblong* is a shape which infantry have at all times taken when obliged in open ground to march in the face of cavalry.—
Though

Though the mode of placing one or two battalions in this manner may be prescribed; yet the various formations of which a greater number are susceptible, depend on ground, the position of the troops, the movements of the enemy, &c. and must be made in consequence of the local orders of the commander: It is therefore from circumstances, and from the flexibility of the military order; that in an instant, he will determine into what shape, the body which he conducts must be thrown.——Should such bodies be at the same time liable to the united attacks of a very superior infantry or artillery; such situation would be critical indeed, and from which nothing but the most determined resistance could extricate them.

Chequered Retreat of the Line

1. *All manœuvres of a Corps retiring*, are infinitely more difficult to be performed with order, than those in advancing.—They must be more or less accomplished by chequered movements; one body by its numbers or position, facing and protecting the retreat of another; and if the enemy presses hard, the whole must probably front in time and await him; as the ground narrows or favours different parts of the corps must double; mouths of defiles and advantageous posts must be possessed; by degrees the different bodies must diminish their front, and throw themselves into column of march when it can be done with safety.

2. *The chequered retreat*, by the alternate battalions or half battalions of a line going to the rear, while the others remain halted, cover them, and in their turn retire in the same manner, is the quickest mode of refusing a part of a corps to the enemy, and at the same time protecting its movement, as long as it continues to be made nearly parallel to the first position.

Fig. 126.

3. If six battalions are in line, the 2d division or the three even ones (2d, 4th, 6th) counting from the right, will go to the right about, retire in line about 200 paces, and then halt front, having carefully preserved their intervals.—The two outward battalions of the retiring ones, will each when it first faces about, form a flank of its outward platoon.—As soon as the second division begins to retreat, all the battalions of the first one will immediately throw back their wing platoons 1-8th of the circle, and thereby when neceffary procure a crofs fire in the intervals, and along the front.

4. When the 2d division fronts, the first one moves up its flanks and is ordered to RETIRE through the intervals, and to form at an equal distance in the rear: As soon as the first division arrives near the second one; that second one begins to *fire* by platoons standing in the same manner as the first hath already done.—The wing platoons of all the second division battalions, place themselves on the flank, as soon as the first division hath paffed them, and remain fo till their turn of retiring is again come.

5. *During the retreat* fhould favourable heights or fituations prefent themfelves to either of the divifions, they fhould be for the time occupied

cupied by the moſt contiguous battalions, who will halt, or incline as is neceſſary, without ſcrupulouſly adhering in that caſe to the alignement, or intervals; and any battalions that may happen to poſſeſs an advanced height, ſhould throw their wings back, and aligne them on their neighbouring battalions, that they may be flanked by ſuch battalions.

6. The retiring diviſion, will move by a directing battalion, and any faults in the halt of the line, can eaſily and muſt be corrected, before the other diviſion arrives at it.

7. The ſecond retiring diviſion having the intervals of the firſt to paſs and to move on, as a guide, can have no difficulty in its movement or direction.

8. During the retreat ſhould any of the flank platoons be ordered to preſerve their flank poſition, ſuch platoons will then march in file.

9. The operation is repeated till the commanding officer halts and fronts a retiring diviſion in the intervals of a ſtanding one.

---

10. *In the courſe of the chequered retreat, a poſition oblique to the original one may taken up.*—The diviſion of the line which is to retire, after facing to the rear, will immediately wheel up its platoons (   ) paces, according to the degree of obliquity which will be required for the line, it will then march on in this echellon poſition, and, when ordered, form, as before directed (S. 157.) to the leading battalion, and front: The other diviſion of the line, when it is to retire, will perform the ſame movement, the battalions forming in their proper intervals, and com-

Fig. 126. B.

pleting

pleting the line as they feverally arrive at it.—Or the retiring divifion, after facing to the rear, will march on, till arrived at the diftance where one flank is to be pofted, it will there halt, and upon that flank inftantly make a change of pofition by the echellon march, and then front: The other divifion will then retire in the fame manner, till it arrives at the pofted one, and it will there inftantly commence its change of pofition, take its place in the intervals, and front in full line.——From this fituation the parallel retreat may continue to be made, or a new oblique direction be again required.

11. In fituations of retreat not very critical, the oblique pofition which the retiring divifion takes up, may be quickeft and eafieft gained by the filing of platoons.

---

12. *In the chequered retreat,* the following rules muft be obferved.—The battalions of the divifion neareft the enemy, will form flanks as foon as there is nothing in their front to cover them; but the other divifions will have no flanks except to the outward battalion of each. The battalions always pafs by their proper intervals, and it is a rule in retiring, that the left of each, fhall always pafs the right of the neighbouring one.——Whatever advantages the ground offers are to be feized, without being too critically tied down to intervals, or to the determined diftance of each retreat.——The divifion next the enemy, muft pafs in front, through the intervals of the divifion immediately behind, and any battalion that finds it neceffary, muft incline for that purpofe.——The retiring divifion muft ftep out, and take up no more time than what is neceffary to avoid confufion.——The divifion neareft the enemy, *fires* by platoons ftanding: the flanks of its battalions only

fire

fire when the enemy attempts to push through the intervals; when that division retires, it fires on skirmishers by men detached from its light company if present, or from platoons formed of rear rank men of one or two of the companies, and placed behind the flanks of the battalions. But should any of its battalions be obliged to halt and to fire, a shorter step must then be taken by the line; and should the enemy threaten to enter at any of its intervals, besides the fire of its flanks, such platoons of the line behind it, as can with safety, must give it support.

———————

13. *If a line with reserves*, finds it necessary to retreat in face of an enemy.———The alternate battalions and the reserves will retire 200 or 300 paces, and then front. The other battalions will then retire; and when they join the first, the reserves also will again march, and front at like distances, the reserves always leading the retreat.———This will continue, till it is proper or safe to break into column of march; the cannon and skirmishers of the whole, covering the front of the retiring line.

———————

14. *Two full lines* will generally make their retreat by passing alternately thro' each other.———Or they may retire by the chequered movement of each line.

15. *If by the chequered movement*; and that the distance between the lines is 300 paces, each will give to its second division 150 paces for its retreat, and thus divide the distance.—When the second division of each line hath retired and fronted; the first division of the first line

line will retire thro' the intervals of its own second division; it will then when it arrives at the first division of the second line pass by files through the battalions of that division; and in the same manner will it pass thro' the intervals of the next division; and when 150 paces in the rear of the whole, the platoons will halt front, and wheel up into line.———The 2d division of the first line having prepared its flanks, begins to retire, as soon as its own first division hath passed the first of the second line: That second division will march thro' the intervals of that first; it will then proceed and pass by files thro' the battalions of the second division of the second line; it will continue in that shape, and pass thro' the intervals of its own 1st division, which is by this time reformed, and when at its 150 paces in the rear of the whole, the platoons will halt front, and wheel up into line.—The divisions of the second line proceed in their turn, exactly in the same manner.

Passage of Lines.

In narrow grounds, where there are redoubled lines, and in many other situations, it becomes necessary for one battalion to PASS directly through another, in marching either to front or rear —But this must particularly happen, when a first line, which has suffered in action, retires through, and makes place for a second line which has come forward to support it;—or—the second line remaining posted, when the

first

first falls back, and retires through it, and thus alternately, till a safe position is attained.

Should the second line be ordered to advance, and occupy the ground which the first is to quit —— As soon as it has approached within 20 paces, and halts, the *front line* battalions receive the word pass to the rear; each platoon is ordered to face to the right, and disengage its head: at the word march, each platoon moves off in file, at a quick step to the rear, and passes straight through the second line.—The officers of the retiring line having been cautioned that they are again to form at 150, or 200 paces from the other line, and having begun to take and count them from the passing of that line, will accordingly be ordered to halt and front; the officers place themselves on the pivot flanks, take their just distances, and aligne to the front of the column, on the three or four first leading platoons, which will be instantly arranged in a true line by a mounted officer.

Passage of a first line, when the second line advances.

Fig. 127. A.

Fig. 127. B.

Wherever the heads of the retreating files present themselves, the officers of the second line cause four files of their platoons to fall back, and again to resume their places when the others have passed.—During the march to the rear, the heads of files must preserve their accurate distances from the left, that when the column halts and fronts, it may immediately be in order to wheel up into line.

If the second line remains posted.—The *first* retires in front, till within 20 paces of the second.—At the word pass by files, each officer turns his platoon to the left—marches quick in file through the second line,

Passage of a first line, when the second line is posted.

and

and halts fronts, at his determined number of paces from that line, by word of command.—The pivots of the column are dreffed, and the platoons wheel up into line.

When a line of feveral battalions hath paffed in this manner, and fronted in column, it is neceffary to drefs their pivots correctly, before wheeling up into line.——The commander of the head battalion will inftantly place the pivots of his three firft platoons in a true direction, and order the officers of his other platoons to line on them, himfelf remaining with the head platoon as the point d'appui, will fee that this is correctly done.——The firft battalion thus fteadied, will become a fufficient direction for the fecond, and every other one to prolong it, by their adjutants; and this operation, though fucceffive from platoon to platoon, and from battalion to battalion, may be performed quickly and correctly; if the adjutants are timeoufly detached, and if the head of the column is quickly arranged.

Re-forming the firft line by a flank battalion.

Fig. 127. B.

Should it be thought proper to give the *alignement from a central* (c), rather than a flank battalion——In this cafe, after halting and fronting, the platoon pivots of the given battalion are from its head accurately lined by its commander, in the true direction.—This battalion being placed, from which diftances and dreffing are taken, the others will inftantly proceed to line their pivot flanks upon it: thofe that are behind it will readily do this; thofe that are before it will find more difficulty, as they muft take their diftances from the rear;—to facilitate this, their platoon officers will face to the directing battalion, and will then fucceffively take their diftances and covering from their then front; as foon as each has acquired his true pofition, he will face about,

Re-forming the firft line on a central battalion.

Fig. 127.

about, and make his platoon join to and drefs to him.—The line will then be ready to form by wheeling up to the pivot flank.

***

Should a new pofition (D), not parallel, be taken by the paffing battalions.——The commander, with his two leading platoons, will firft enter it, and direct the others to regulate their flanks by them; and, if feveral battalions are paffing the fecond line, the NEW alignement is thus made eafier for them.

<small>Firft line reforms in an oblique pofition. Fig. 127.</small>

***

When a height (E), in the rear is to be *crowned* by a retiring line.— Each officer muft not drefs exactly to the platoon that precedes him, but in joining it, he muft *halt*, and arrange his own in fuch a manner, that the SLOPE of the rifing can be entirely feen and commanded, which is here the great object, and would not be attained if the troops were to adhere to a ftraight line.

<small>Crowning a height. Fig. 127.</small>

***

A line which hath paffed, will often before forming, throw back a *wing*,—in order to occupy a particular pofition,—to prevent the enemy's defigns on that wing; or at leaft to make him take a greater detour to effect it,—or—that he may be obliged to aligne his own on a height which is occupied, and from which he may be flanked.

When the movement is forefeen, and according to the wing which is to be refufed, fhould the platoons of the line pafs.——If the left is to be pofted, and the right **refufed**, the platoons may pafs from their *left*, the column will thereby have its left in front, will be more readily

dily directed on the point d'appui, and the preservation of distances will be facilitated, as they will then be taken from the front. If the right is to be posted, the platoons may pass from their right.

Refusing a wing.

The line (A B), is here supposed to *refuse* its four right battalions (D), after having retired and passed as a column with its left in front.—— All the battalions of the line to the left of the fourth, proceed as already directed, and take up their line from the left or head battalion, which, in the course of the movement, is conducted to the point d'appui, where its head is to rest, and from whence the direction of formation is to be given; but the left of the fourth battalion becomes a new point d'appui, upon which the four battalions of the right are thrown back into any situation which the nature of the ground, and the views of the commander may require.

Fig. 128.

It may happen where the *passing* line is to *post* one flank, and *refuse* the other, that the officers will have their distances to take from behind; the original remedy for this inconvenience has been shewn, another also may be readily applied, which is to halt the whole, at any time after passing, and to counter-march each platoon, which will then cause the future formation to be taken from the front of the column.

A retiring line may also *refuse* a wing, by forming in line very soon after passing, and then taking up an oblique position to the rear, by the echellon march, or some other of the modes already prescribed.

OF SECOND LINES.

1. No confiderable body fhould ever be formed, without a proportion of it being placed in *referve* or in *fecond line*, and more or lefs ftrong, according to circumftances.—The movements of fuch fecond line will always correfpond to thofe of the firft, and it will always preferve its parallelifm, and diftance.—If the firft line makes a flank, or central change of pofition, the fecond muft make a change alfo on fuch point, as will bring it into its relative fituation.

2. The march of the fecond line in front, is regulated by its own divifion or battalion of direction, which moves relatively to that of the firft line.——In forming in line it will march upon its own points which are parallel to and afcertained in confequence of thofe of the firft.

3. When the lines break in columns to the front; the fecond will generally follow thofe of the firft.—When the march is to the flanks; the fecond line will compofe a feparate column, or columns.—When the march is to the rear; the fecond line will lead in columns.

4. The diftance betwixt the lines, may be in general fuppofed equal to the front of two battalions, and an interval,

5. Second lines are feldom compofed of as many battalions as the firft; they are often divided into diftinct bodies, covering feparate parts of the firft line.

6. Second lines will not always remain extended, they will often be formed in column of battalions, or of greater numbers, ready to be moved to any point where their affiftance is neceffary.

7. Whenever the firft line breaks and manœuvres by its right to face to the left, or by its left to face to the right.—The movements of the fecond line are free and unembaraffed, and it may turn round the manœuvring flank of the firft line, and take its new pofition behind it, by extending itfelf parallel to that direction, how oblique fo. ever it may be.

8. The central movement generally required from the fecond line to conform to that of the firft, is equivalent to that line marching in two columns of platoons, from near the center obliquely to the front, and from that fituation forming to both flanks.

9. The movements of the central columns being well underftood.— Thofe of the battalions of the wings, are fimilar in the two lines.

10. The officer commanding the fecond line, muft always be properly informed of the nature of the change to be made by the firft, that he may readily determine his correfponding movements.

11. It requires much attention—To conduct heads of battalion columns of both lines nearly parallel to their lateral ones, and perpendicularly or diagonally to front or rear, according to the nature of the movement.—To determine with precifion, and in due time, their points in the new line, that wavering and uncertainty of march may be avoided.—In great movements to allow the foldier every facility of
motion,

motion, without encreasing the distances of divisions, and to require the most exact attention on entering the new line and in forming.—To avoid obstacles in the course of marching, but as soon as possible to re-enter the proper path of the column.—While out of that path, the colours of that battalion column may be lowered, (as a mark for the neighbouring column, not to be then entirely regulated by it,) and again advanced when it regains its proper situation.

12. In many cases, and where great concert of movement is not required, a second line may form battalion columns at half distance, each behind its flank nearest to the new position, and relative points being prepared, each will march up, and prolong the line.

13. All the battalions of a second line, must at the completion of every change of position, find themselves placed in the same relative situation with respect to the first line, as they were in before the commencement of the movement.

14. All changes of position of a first line are made according to one of the modes already prescribed; in general in critical situations they are made on a fixed flank, or central point, and by the echellon march of platoons.—But the movements of a second line being protected, more complicated, and embracing more ground, are made by the march of battalion columns regulated by a certain determined division of the line.

---

15. In all cases where a change of position is made on a flank or central point of the *first line*, the movement of its *corresponding* point of

the

the second line determines the new relative situation of that second line.

16. *To find this point*, it is necessary to premise, that if a circle is described from any point A. of a first line AE with a radius equal to the distance betwixt the two lines; then its corresponding point a, in the second line will be always in the circumference of that circle, at such place as the second line becomes a tangent to the circle.—Should the first line therefore make a change of position AR either on a flank or central point A; its *corresponding* point a, at that time in the second line, will move so as still to preserve and halt in its relative situation a, 2; and by the movement and halt of that part, preceded by the one d, of *interfection*, every other part of the second line, either by following them, or by yielding from them is regulated, and directed. —Betwixt the old and new situation of the corresponding point, a, and equi-distant from each, lies the point d, where the old and new positions of the second line *interfect*, and which is a most material one in the movement of that line.

Fig. 117-129.

S. 207. *When two Lines change Position on a central point of the first Line.*

1. A. is the point on which the change is to be made; a. is obviously its corresponding point in the second line whose distance in paces is known.—The direction of the first line AR, being ascertained, it becomes immediately necessary to mark the *corresponding* point a, 2. in the second line, and also the point of *interfection* d, that the prolongation of that line may be also determined.—From the point A, therefore

Fig. 129.

fore and in a direction perpendicular to the new line, a person accurately paces the known distance between the two lines, and halts at a, 2; and from thence observes by his eye, the perpendicular to the line which he has just paced, and also its intersection with the second line, which gives the platoon or point d, together with the direction of the new second line.——The points a, 2. and d, being thus fixed, the lines proceed to make their movement, viz. the first line by the echellon march (S. 195.); the second line B breaks inward to the platoon d; that platoon moves its pivot flank along the new line followed by all those betwixt it, and a, till a, arrives at the point a, 2.—The other platoons and battalions of the second line move relatively to the part a, d, the whole performing the precise operation already detailed in the change of position of the line on a moveable central point S. 185.

2. But in order to accelerate the movement of the second line in *central* changes; a platoon or flank point o, as much beyond a, as a, is removed from d, may be taken; this point is evidently the one, which will rest at the point of intersection d, when a, is arrived and halted at a, 2.——As soon therefore as the points o, d, are ascertained and without waiting for the progressive movement of this center part of the line, every thing that is in rear of o, may march and form in the new position, regarding o, 2, as its leading flank point; and every thing that fronts to d, regarding d, 2, as its leading flank point will march and form upon it, in the new line accordingly. S. 186. *Fig. 129.*

3. In order the better to ascertain the parallel direction of the new second line, 2 persons separated from each other about 100 paces, may set out from different points of the new first line, and accurately pace *Fig. 129.*

the

the known distance of the second; when they halt, the line of their prolongation gives the new direction, and also the intersection of the second lines.——Or if the first line points on any very distant object the second line from the point a, 2, will be readily judged to pass a very little behind that object.

─────────

S. 208. *When two Lines change Position forward, on a flank of the first Line.*

Fig. 130.

The direction of the first line being ascertained, that line will march into it by the echellon march, S. 195. In the mean time the corresponding flank point a, 2, in the new second line having been taken, and also the point d, in the intersection of the two lines; these points serve as the base of formation.——The second line will break into open column facing to the platoon d, of intersection; the whole will proceed as in S. 185. that platoon marching along the new line till a, arrives at a, 2: and when the other battalions which have moved forward are anew arranged in open column, they wheel up into line.

─────────

S. 209. *When two Lines change position backward, on a flank of the first Line.*

Fig. 131.

The direction of the first line being ascertained, that line will march into it, by the echellon movement to the rear, S. 195.—The corresponding points a, 2, and d, in the new second line serving as a base of formation, having been ascertained as well as the point o, that line breaks into open column facing to the point of intersection: The part

of

of the line between a, and o, marches on to the point of interfection, and from thence prolongs the new lines: The part of the line behind o, regarding d, as its leading flank point of formation, will march and form upon it, in the new line accordingly. S. 186.

S. 210. *When two formed Lines, wheel into open Column, march to a flank, change direction, and take up a new pofition.*

1. *If the new pofition is a retired one.*—Both lines wheel into open column (fuppofe the left in front) and move on.—C, is a point where the head of the firft line is to change its direction into that of C, D, Fig. 132, by wheeling on its pivot flank: The leader of the fecond line being apprized of this point fends forward to afcertain his correfponding point c, and his parallel direction c, d; thefe two points become the bafe of formation for the fecond line, and d, is in the interfection of the old and new lines.—Both columns proceed in their firft direction, and when the firft line arrives at C, the head wheels on its pivot into the direction C, D, followed by the reft of its column; but the fecond line then changing its head on the point c, moves towards it, there to enter its correfponding direction.—The firft line halts, and fuch parts of it as are in the new direction remain fo; while the rear of the column, by the movement of S. 182. gains the new direction, and by wheeling up forms in line;—or, if the ground permits, it enters the new line by the echellon march of S. 158; each divifion firft counter-marching by files, and then facing about, fo that its rear rank may lead.

2. When the first line halts; if the head division of the second line *has not entered* the new direction, it still moves on to its point of entry c, prolongs the line till it arrives behind its corresponding point A, 2. of the first line, it then halts, and also all such others as have arrived in the new direction; while the rear of the column, which since the halt of the first line has been gradually (by obliquing) and regularly throwing itself to the left, places itself in open column on the new direction and wheels up into line.

3. When the first line halts; if the head of the second *has entered* the direction, it moves on till it arrives behind its corresponding point of the first one, and the rear of the column obliques to avoid interfering with, and to make place for the forming of the first line.—When the head halts, such part of the column as is not in the new direction gains it by the movement of S. 182.

4. If the head of the first line when it arrives at E, *waits* till the 2d line arrives at the corresponding point c, they will then proceed equally.——Or, the march of the second line may begin proportionally sooner, than that of the first.

———————

Fig. 133.

5. *If the new position is an advanced one.*—The first line changes its direction at C, by a wheel to the reverse hand into C D; marches on till the head is halted; such part of the column as is not in the new line enters it, by the divisions of the column wheeling back into echellon, and then marching up successively into line.—The second line which has ascertained its corresponding points c, d, moves on, enters the new direction at d, by a wheel to its reverse hand, and prolongs

the

the line till it arrives at its flank point, when the whole halt, and the battalions of such part of the column as are not then in the new direction gain it by the movements of *S.* 182.

6. The second line must take care not to pass its point d, but after arriving at it must wait till the head of the first line, which has more ground to go over, arrives equal with it in the new line; it will then move on.

Of the Column of Route.

1. The column of route formed by divisions* of the battalion is the foundation of all great distant movements, and even of evolutions and manœuvres.—It is in that order that the battalion should at any time be permitted to move; that the columns of an army should perform their marches; that an enemy should be approached; and that safety can be insured to the troops in their transitions from one point to another.—All marches are therefore made in column of divisions of the line, and never on a less front than 6 files where the formation is 3 deep, or 4 files where it is 2 deep, nor does any advantage arise from such column, if it is an open column, exceeding 16 or 18 file in front, where a considerable space is to be gone over.

*By companies, or other divisions.

2. At no time whatever ought a column of manœuvre or of route to occupy a greater extent of ground in marching, than what is equal

to its front when in order of battle; no situation can require it as an advantage.—Therefore the marching of great bodies in file, where improper extension is unavoidable, must be looked upon as an unmilitary practice, and only to be had recourse to, when unavoidably necessary.—Where woods, inclosures, and bad or narrow routes absolutely require a march in file, there is no remedy for the delay in forming, and man may be obliged to come up after man; but these circumstances which should be regarded as exceptions from the primary and desired order of march on a greater front, should tend the more to enforce the great principle of preventing improper distances, and of getting out of so weak a situation, as soon as the nature of the ground will allow of the front of the march being encreased.

Disadvantages of file marching.

3. In common route marching, the battalion or more considerable column may be carried on at a natural pace of about 75 steps in a minute; or near two miles and a half in an hour: The attention of the soldier is allowed to be relaxed, he moves without the restraint of cadence of step, or carried arms; rear ranks are opened to one or two paces; files are loosened but never confounded; in no situation is the ordered distance between divisions ever to be encreased, and the proper flank officers and under officers remain answerable for them.

Rate, and circumstances of march.

4. If the column is halted, the whole must be put in march at the same time.—The movement of the head division must be steady and equal; the descending of heights must not be hurried, that the part of the column ascending may properly keep up.—Alterations occasioned by the windings of the route are executed without losing distance.—Soldiers are not to break to avoid mud or small spots of water.—

Attentions in march.

The pivots muſt trace out ſuch a path for themſelves as will beſt avoid ſmall obſtructions, and the men of the diviſions will open from, and not preſs upon their pivots.—When platoon officers are permitted to be mounted, each will remain on the flank of his diviſion watching over its exactneſs, and that the proper diſtance of march is kept by the flank pivot under officer appointed to preſerve it.

5. Where the arrival of a column at a given point is to be perfectly punctual, in that caſe the diſtance being known, the head muſt move at an equal cadenced ſtep, and the rear muſt conform; and a perſon expreſsly appointed will at the head of the column take ſuch ſtep as the nature of the route ſhall permit the column to comply with.

6. Nothing ſo much fatigues troops in a conſiderable column, and is more to be avoided than an inequality of march.—One great reaſon is, that the rear of the column frequently, and unneceſſarily deviates from the line which its head traces out; and in endeavouring to regain that line and their firſt diſtances the diviſions muſt of courſe run or ſtop and again take up their march.——It is unneceſſary to attempt the ſame ſcrupulous obſervances in common route marching, as when going to enter into the alignement; but even a general attention to this circumſtance will in that caſe prevent unneceſſary winding in the march, which tends to prolong it, and to harraſs the ſoldier.

7. When the probable required formation of the line will be to a flank, then the column of march is an open one, and except the cannon no impediment or circumſtance whatever muſt be allowed betwixt the diviſions or in the intervals of battalions.—When cannon can poſſibly

bly move on the flank of the battalion they ought, and mounted officers or bat-horses must not be permitted betwixt the divisions.—If the probable formation may be to the front, then distances are more closed up, and bat-horses, &c. may be allowed, betwixt the brigades of a column, but not betwixt the battalions of a brigade.

8. It is always time well employed to halt the head of a considerable column, and enlarge an opening or repair a bad step in the road, rather than to diminish the front, or lengthen out the line of march.—No individual is to presume to march on a less front than what the leader of the column directs, and all doublings must therefore come from the head only.—The preservation of the original front of march on all occasions, is a point of the highest consequence, and it is a most meritorious service in any officer to prevent all unnecessary doublings, or to correct them as soon as made; no advantage can arise from them, and therefore each commanding officer, when he arrives near the cause, should be assured that it is necessary before he permits his battalion so to double: On all occasions he should continue his march on the greatest front that without crowding, the road or overtures will allow, although the regiment or divisions before him may be marching on a narrower front.

Overtures of march.

9. All openings made for the march of a column should be sufficient for the greatest front on which it is to march, they should all be of the same width; otherwise each smaller one becomes a defile.

Avoiding of difficulties in the march.

10. At all points of encreasing or diminishing the front of the march an intelligent officer per battalion or brigade should be stationed to see that

that it is performed with celerity; and the commandant of a considerable column should have constant reports and inspections made that the column is moving with proper regularity: he should have officers in advance to apprize him of difficulties to be avoided or obstacles to be passed, and should himself apply every proper means to obviate such as may occur in the march. (And at no time are such helps more necessary than when regiments are acting in line on broken ground, and when their movements are combined with those of others.)—When the column arrives near its object of formation or manœuvre, the strictest attention of officers and men is to be resumed, and each individual is to be at his post.

11. The great principle on all occasions of diminishing or encreasing the front of the column in march is—That such part as doubles or forms up, shall flacken or quicken its pace, as is necessary to conform to the part which has no such operation to perform, but which continues its uniform march, without the least alteration, as if no such process was going on; and if this is observed, distances can never be lost, or the column lengthened out.——Unless the unremitting attention and intelligence of officers commanding battalions and their divisions are given to this object, disorder and constant stops and runs take place in the column; the soldier is improperly and unnecessarily harrassed; disease soon gains ground in a corps thus ill conducted, which is not to be depended on in any combined arrangement, is unequal to any effort when its exertion may be required, and is soon ruined from a neglect of the first and most important of military duties.

12. The most important exercise that troops can attend to is the march in column of route.——No calculation can be made on columns which

Importance of exactness in the march.

which do not move with an afcertained regularity and great fatigue arifes to the foldier: A general cannot depend on execution, and therefore can make no combination of time or diftance in the arrival of columns at their feveral points: In many fituations an improper extended column will be liable to be beat in detail, and before it can be formed.—Troops that are feldom affembled for the manœuvres of war can hardly feel the neceffity of the modes in which a confiderable body of infantry muft march and move.

<small>Columns and their diftances.</small>
13. The diftance of columns from each other during a march, depends on the circumftances of ground and the object of that march with regard to future formations.——The more columns in which a confiderable corps marches, the lefs extent in depth will it take up; the lefs frequent will be its halts; and the more fpeedily can it form in order of battle to the front.

<small>Combinations of march.</small>
14. On the combinations of march, and on their execution by the component parts of the body, does the fuccefs of every military operation or enterprize depend.—To fulfil the intentions of the chief every concurrent exertion of the fubordinate officer is *required*, and the beft calculated difpofitions founded on local knowledge, muft fail if there is a want of that punctuality of execution, which every general muft truft to, and has a right to expect from the leaders of his columns.

<small>Nature of marches.</small>
15. The compofition of the columns of an army muft always depend on the nature of the country, and the objects of the movement. Marches made *parallel* to the front of the enemy, will generally be performed by the lines on which the army is encamped; each marching by

by its flank, and occupying when in march, the same extent of ground as when formed in line.—Marches made *perpendicular* to the front of the enemy either advancing, or retiring will be covered by strong van or rear guards.—The columns will be formed of considerable divisions of the army, each generally composed both of cavalry and infantry; they will move at half or quarter distance, and the nature of the country will determine which arm precedes.

16. During a march to the *front*, the separation of the heads of the columns, must unavoidably be considerable; but when they approach the enemy, they must be so regulated and directed, as to be able to occupy the intermediate spaces, if required to form in line.—Some one column must determine the relative situation of the others, and divisions must be more closed up, than in a march to a flank, and in proportion as they draw near to the enemy must exactness and attention encrease. The general in consequence of the observations he has made will determine on his disposition; the columns which are now probably halted and collected will be subdivided and multiplied; each body will be directed on its point of formation and the component parts of each will in due time disengage from the general column, and form in line.

General objects in marches to the front.

17. The safety of marches to the *rear* must depend on particular dispositions, on strong covering rear grounds, and on the judicious choice of such posts as will check the pursuit of the enemy.—In these marches to front or rear, the divisions of the second line generally follow or lead those of the first, and all their formations are relative thereto. The heavy artillery and carriages of an army form a particular object of every march, and must be directed according to

General objects in marches to the rear.

the circumstances of the day.—The safety of the march by the arrangement of detachments and posts to cover the front, rear, or flanks of the columns, depends also on many local and temporary reasons, but are an essential part of the general disposition.

General Remarks.

1. All these OPERATIONS in Line, Column, or Echellon, are applied according to circumstances.—As *Countermovements of Defence.*—*Movements of previous Formation.*—*Movements of Attack*—as well as *Movements of Retreat.*

Advantage of offensive movements.

2. Where there is equal skill in the execution of movements, the *defensive* alteration of position, is sooner made than the *offensive* one, as much less ground is gone over to oppose, than to attack; the great advantage however attending the latter is, that the measure being previously determined on, every thing is prepared for rapid execution before the design is obvious; whereas the counter-movement depending on the appearance of the moment, requires quick observation, immediate decision, instant arrangement, and a disposition simple and that cannot produce hurry or confusion in the execution.——It is in these situations that the justness of distances and of the march in columns allows of decisive operations, which durst not be attempted, unless the moving body could be depended on, as ready at every instant to form up in compleat order.—Manœuvre will chiefly operate

where

where an enemy is inferior in number, unexpert in movement, weakly posted, and where the weak point is found out and attacked before he can move to strengthen it.

3. If the flank of one body is thrown forward, by the same means may that of the other be thrown back.—If one body prolongs its line to outflank, the other may by the same movement maintain its relative situation.—Whatever change of position is made by one body, the other may counteract it by a similar change.—If the wing of one body is refused, the wing of the other may be advanced to seize an advantage. *Counter-movements of defence.*

---

4. A body of troops which has a considerable march to make, previous to the *attack*, must always approach an enemy in one or more columns, at open or other distances according to circumstances.—Some general knowledge of an enemy's situation, determines the manner, in which he is approached, the composition of the columns, the flank of each which leads, and their combination in forming.—A nearer view determines a perseverance in the first direction, or a change in the leading flanks, and direction of the columns, in order to form in the speediest and most advantageous manner. *Movements of previous formation.*

5. *Original Positions* are taken up from the connected movements of columns of march, and entered upon in some of the modes prescribed. —Different feints are used, to prevent an early knowledge of the position intended to be taken, or the point to be attacked, and light troops, cavalry and artillery where the ground, &c. allows, cover all movements of the infantry.—Such original position is either *parallel*, *Original positions.*

or

or *oblique* to that of the enemy posted, and is often changed previous to the attack.——When a considerable corps of troops is to act offensively, it must form in line at latest within 1200, or 1500 paces of a posted enemy, unless the ground particularly favours, and covers from the fire of his artillery; the enfilade of which is what chiefly prevents bodies in column from approaching nearer, and that space under the unceasing fire of their own artillery, troops in line will march over, in 18 minutes.—However quickly columns could move up close to an enemy, yet as they must then form in line, no time would be gained, and their loss be heavier, than when the original formation is made at a due distance.

Movements of attack.

6. From *parallel* position, the attack is made either in line, or by a flank of the line in echellon, that flank being reinforced, and the other refused: or, from a new and advantageous position taken up and not provided against by the enemy.——From *oblique* position the attack is directed against a comparatively weak point of the enemy.—Attacks from the *center* are more liable to enfilade, and sooner guarded against, than from the flank.—It is generally wished to post one wing, and refuse it, and to make the attack with the other reinforced by detached corps, to which the whole strength of the second line is endeavoured to be added; and for these purposes the movement in *echellon;* the *change of position* which gains the flank of the enemy; the *march* on one or more lines; and the *passage* of lines, when redoubled ones are destined to replace each other are particularly applied.

Movements of retreat.

7. General movements of *Retreat* executed by an army must be considered as combinations of columns of march, covered by positions,

and

and a strong rear guard.—Troops are occasionally taken out of the retiring columns of march to occupy positions, and heights; they remain till the rear has passed, and then become the rear guard; this they continue to be, till they find other troops in like manner posted; these last in their turn become also the rear guard, and in this way are the troops of columns in such situations relieved.——A rear guard, will fall back by the *retreat in line*—the *chequered retreat*—the *passage of lines*—the *echellon* changes of position.

8. When a considerable line formed in front of an enemy must retire, or relinquish an attack made, or intended: One wing ought to be originally so posted, that the other by some of the above movements can fall back upon it, and take a new position; being protected in the operation by the enfilade of the posted wing, which in its turn, can fall back upon the other.——The mixed considerations, and support of CAVALRY, INFANTRY, and ARTILLERY on such occasions, require a very intricate discussion.

D. D.

End of FOURTH PART.

MILITARY REGULATIONS.

Directions to the Bookbinder.

To be bound in One Volume, in the following Order:

Title Page, and Adjutant General's Order.

Introductory Preface.

Contents of the Four Component Parts.

Parts { First. Second. Third. }

Inspection or Review———Light Infantry.

Part———Fourth.

The Plates in their proper Order, from 1. to 16.

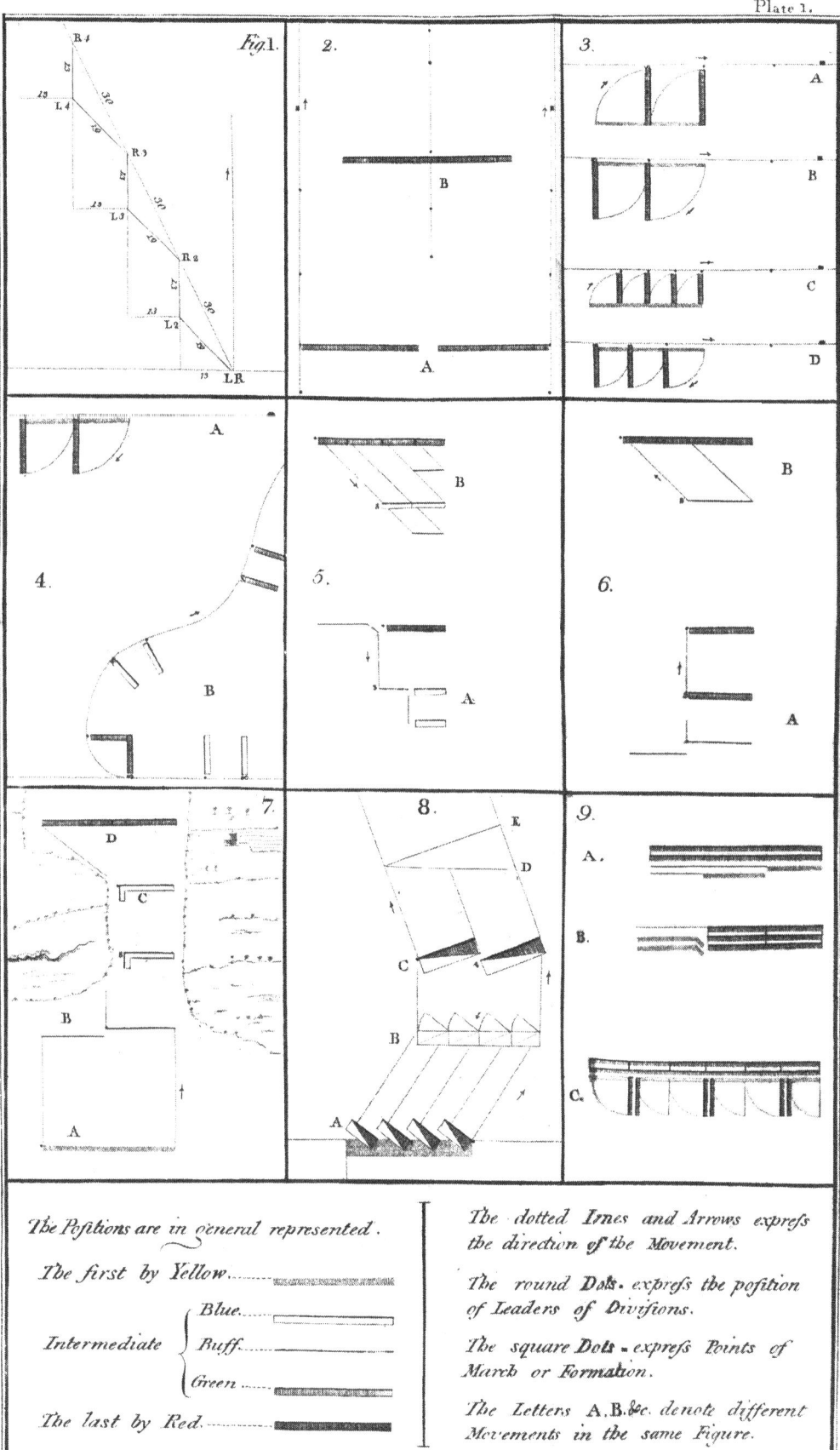

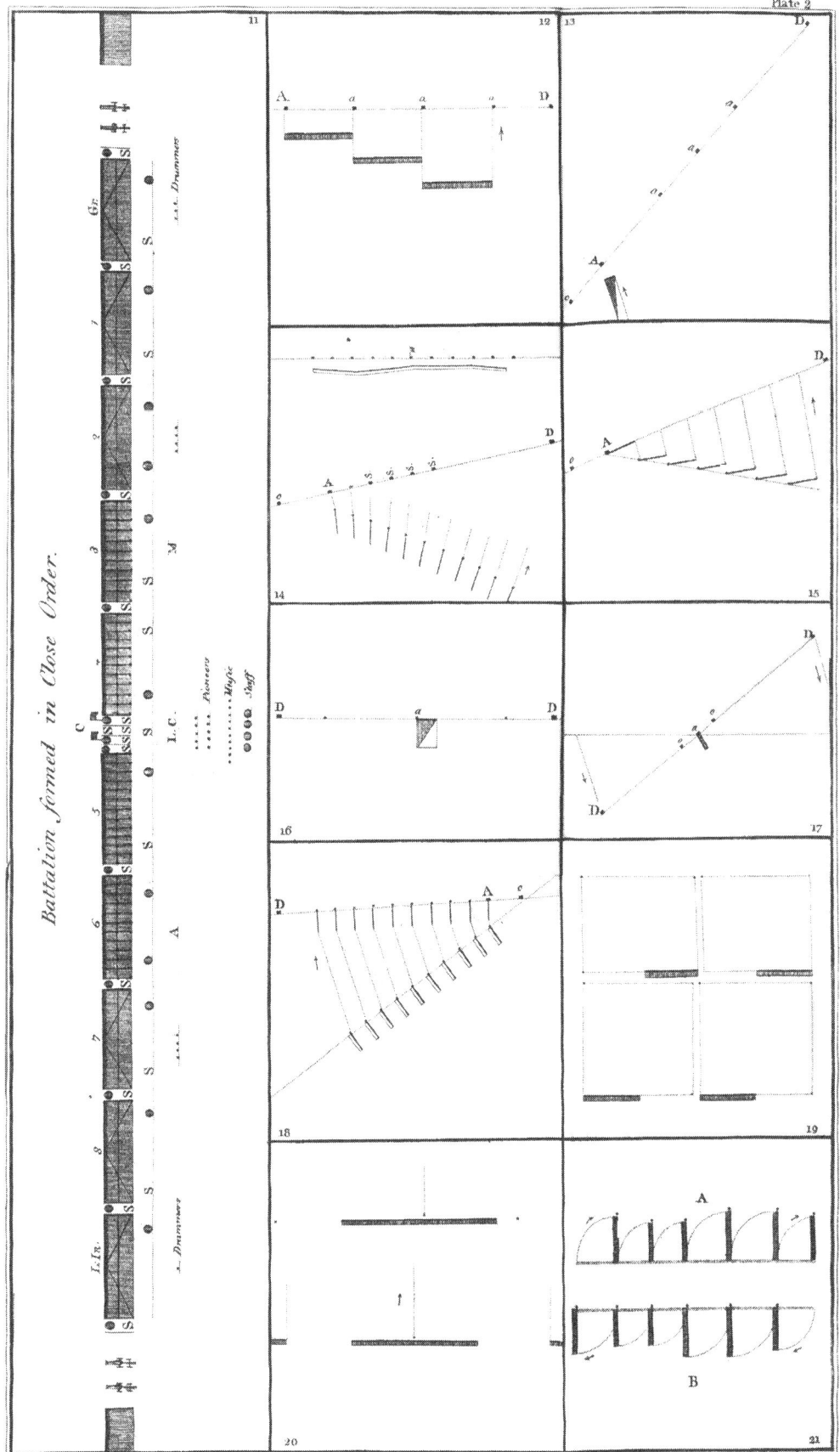

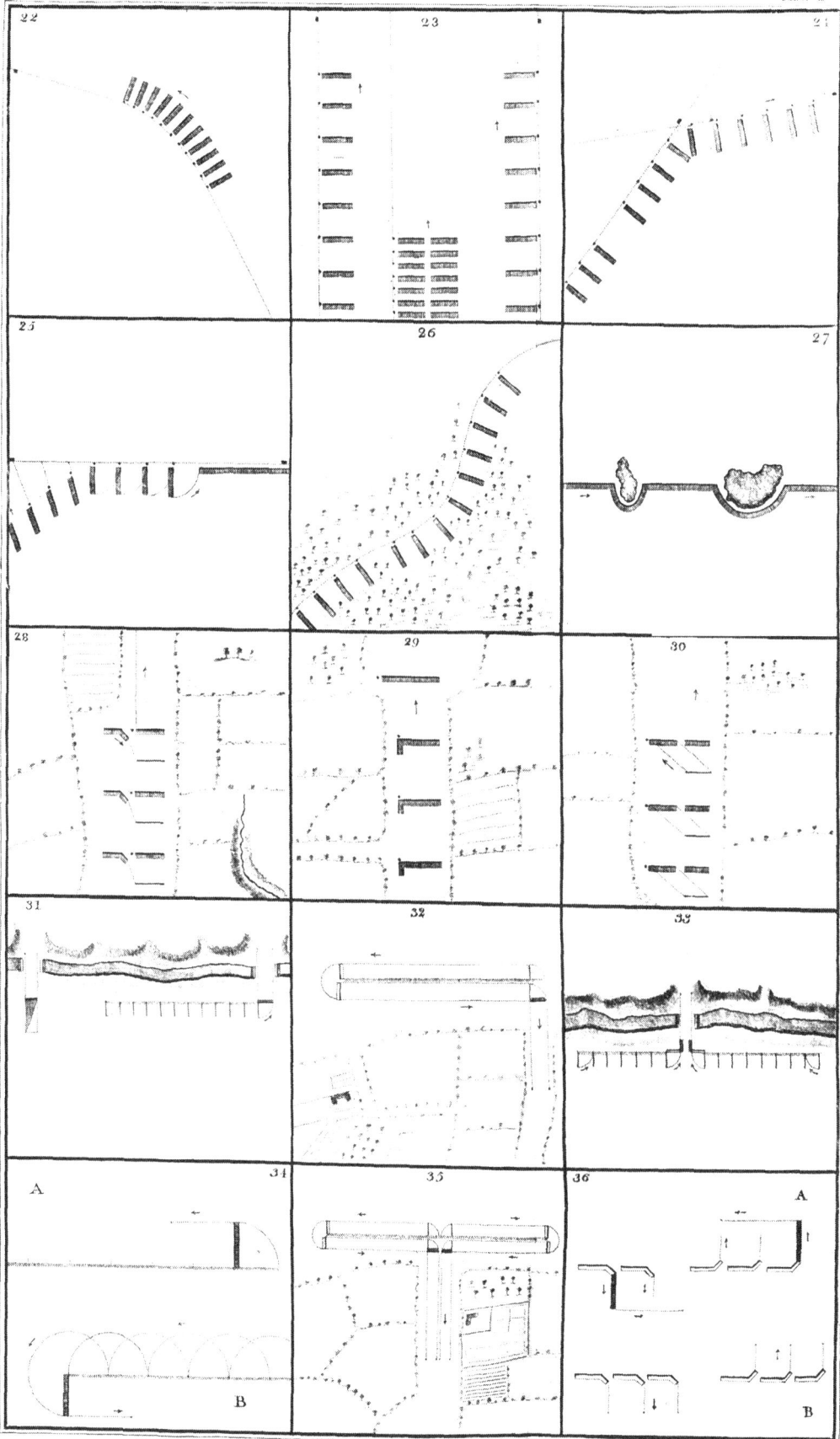

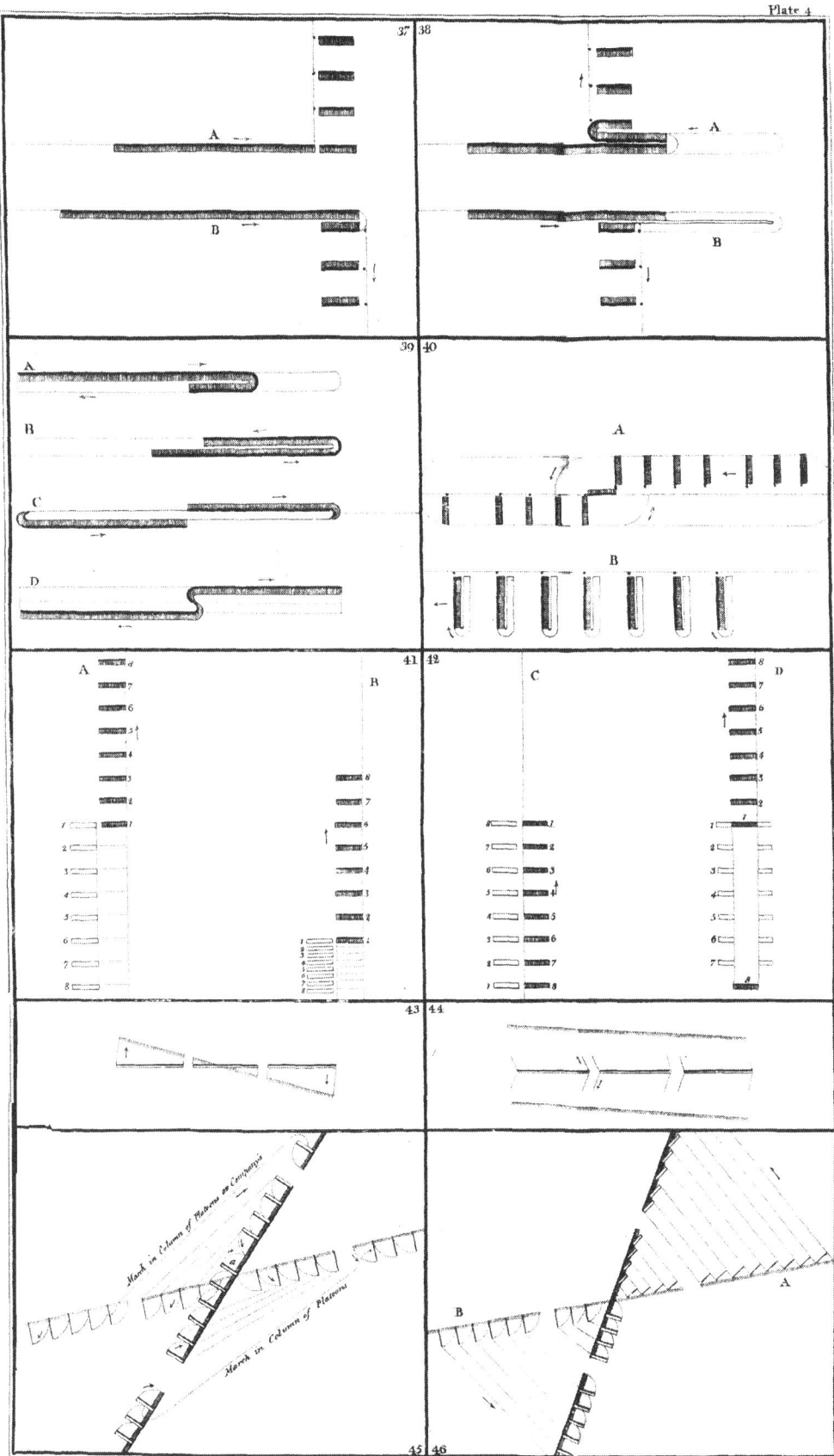

Plate 4

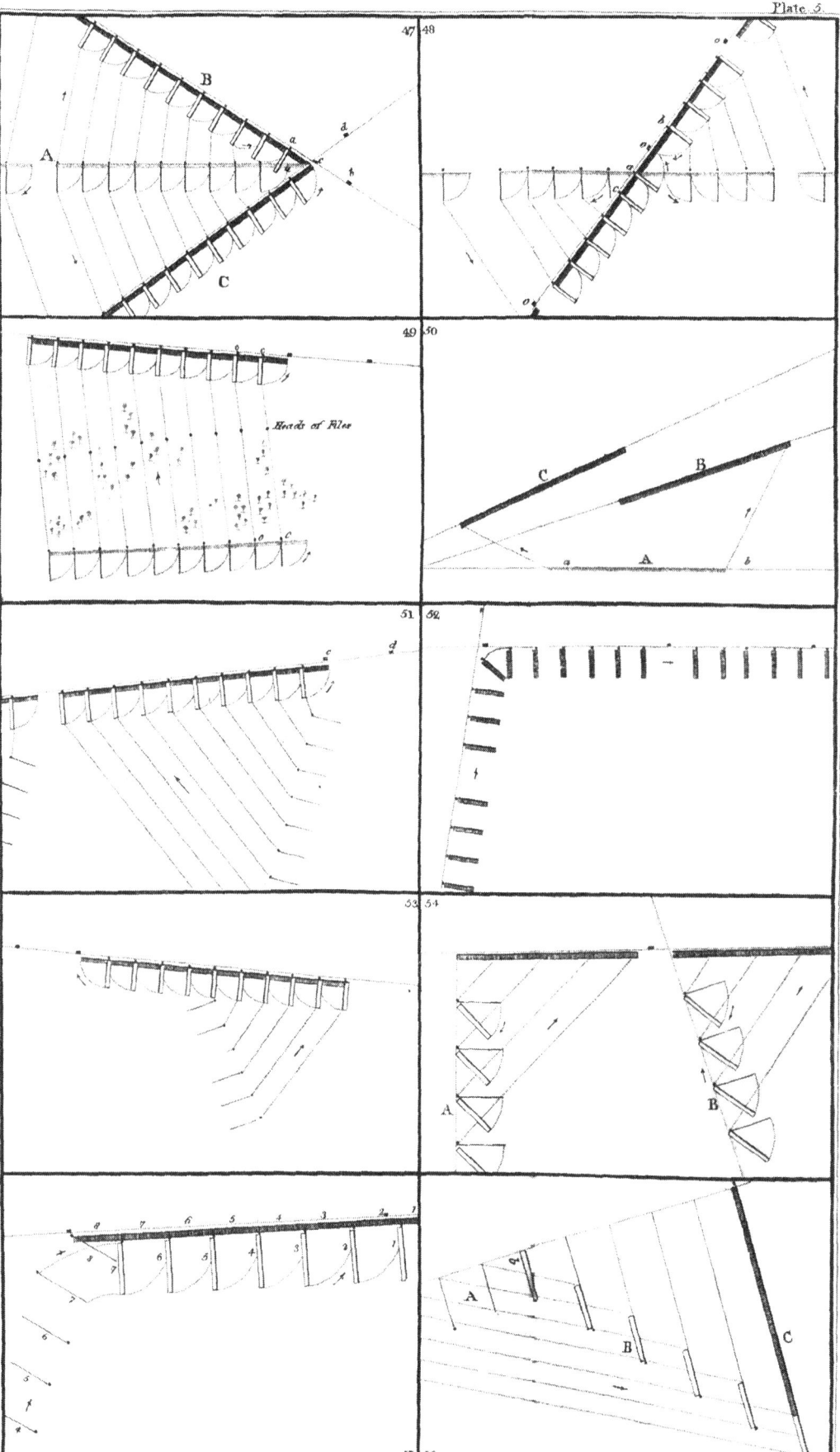

Plate 6.

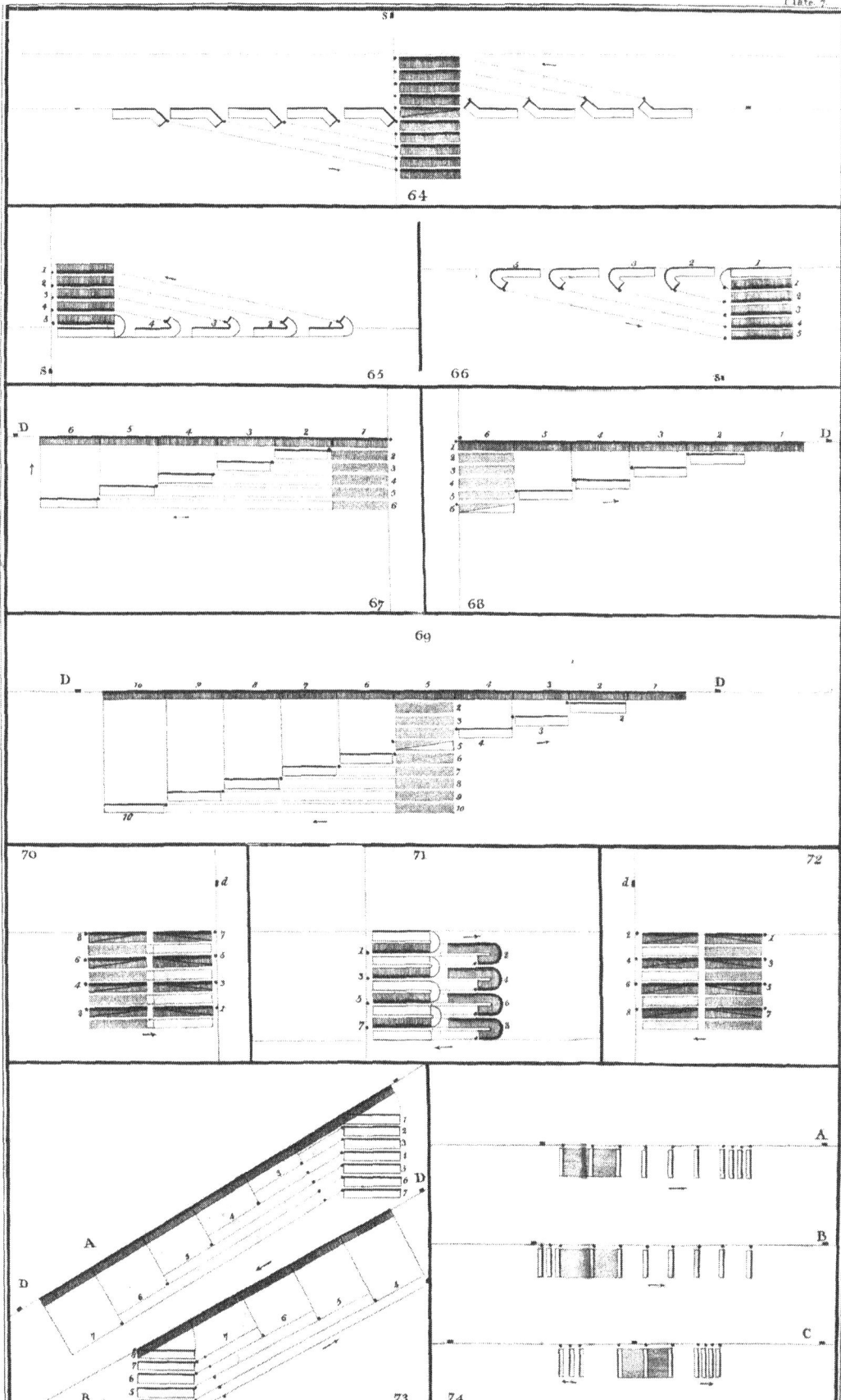

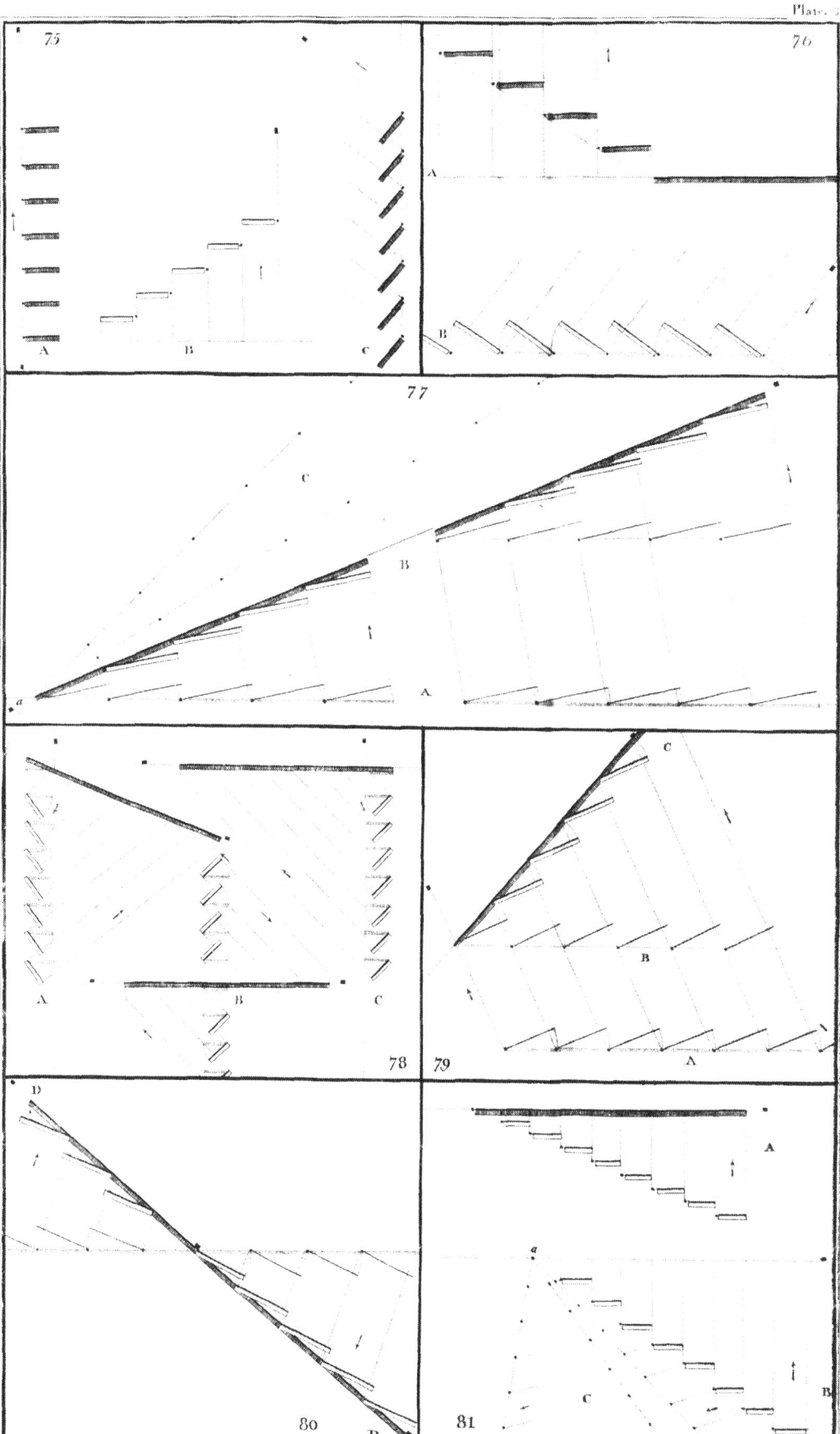

Plate. 9.

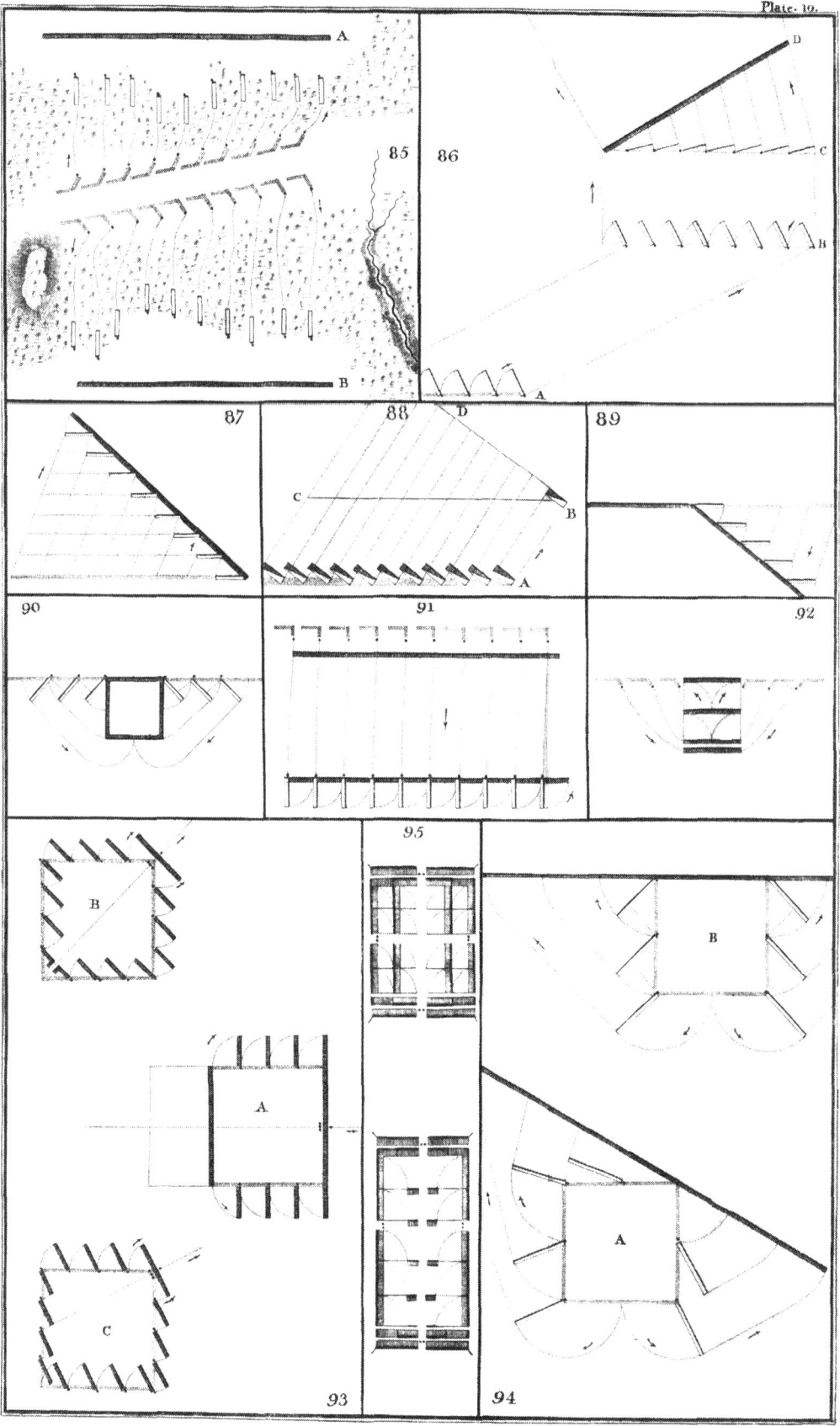

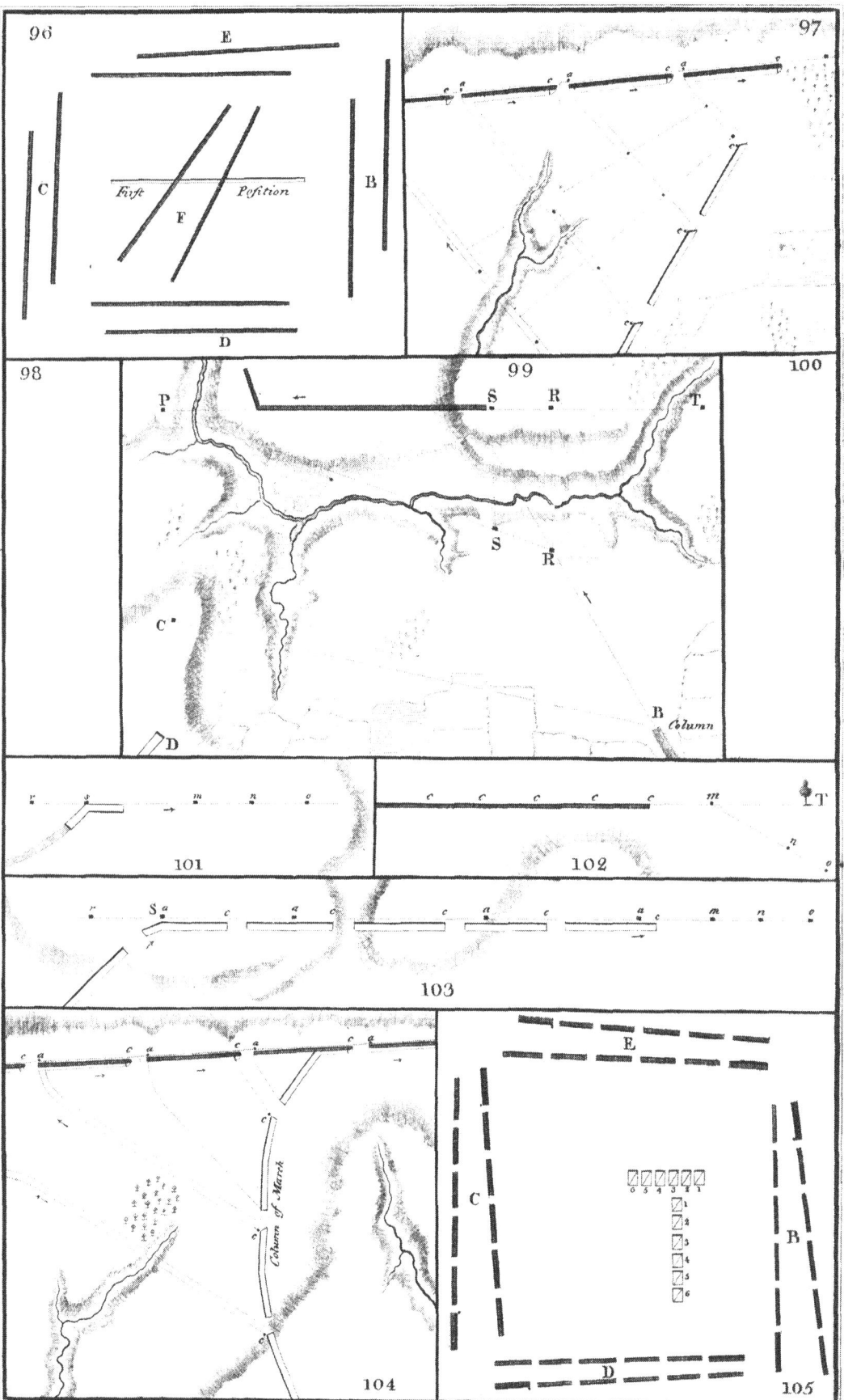

Plate 11

Plate. 12.

Plate. 13.

Plate 14.

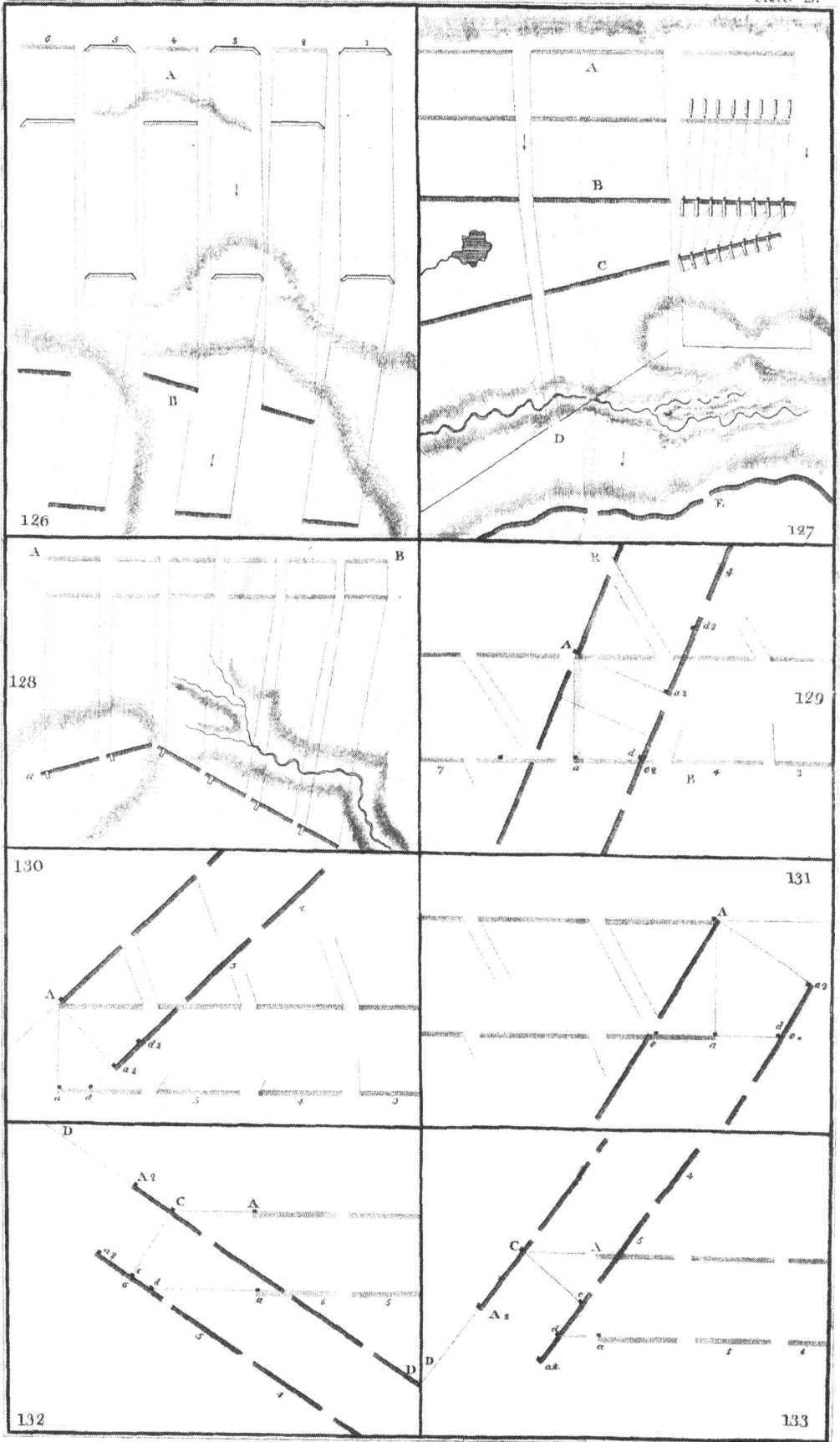

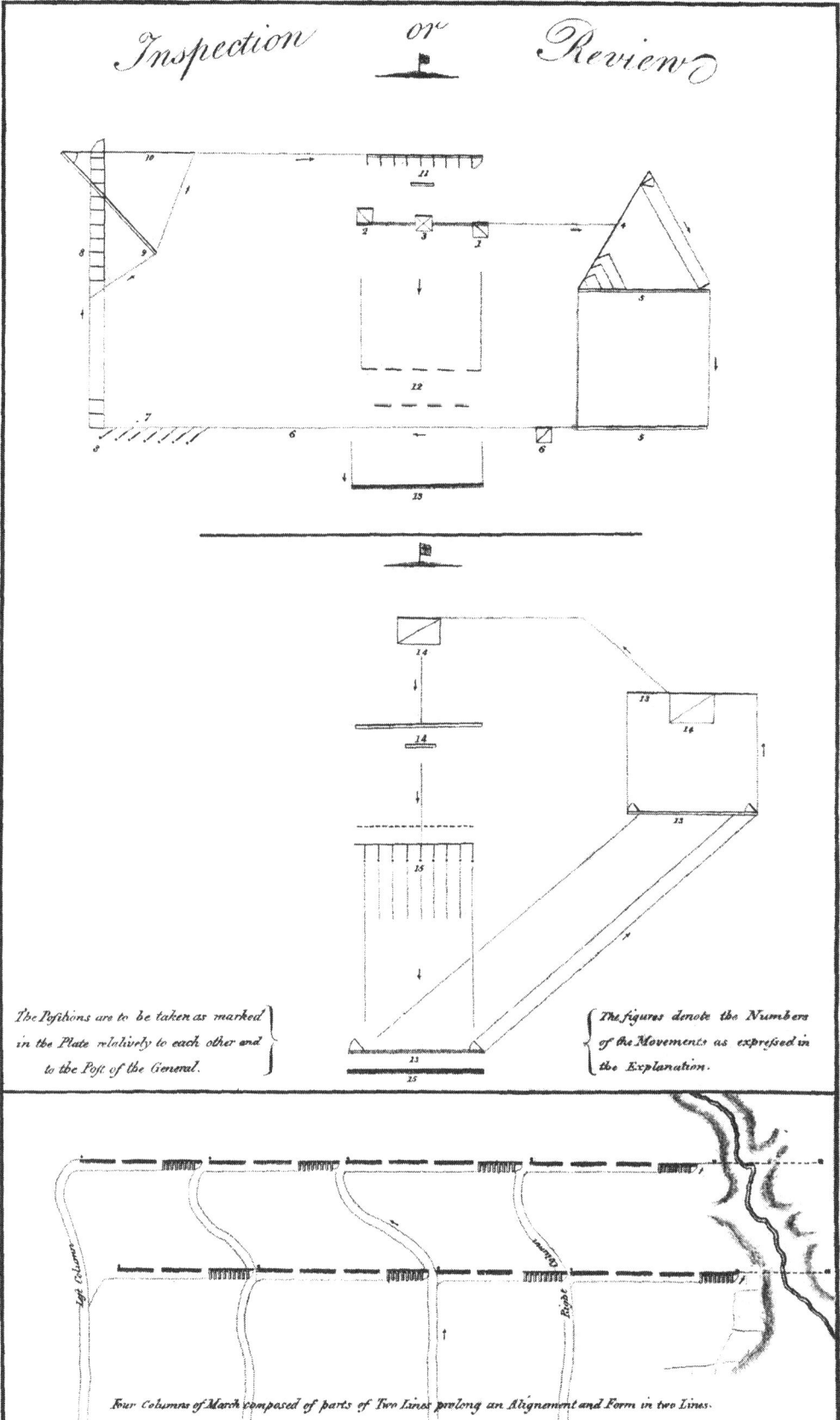

www.ingramcontent.com/pod-product-compliance
Lightning Source LLC
Chambersburg PA
CBHW082032230426
43670CB00016B/2636